Christians in the Digital Age

JOHN COBLENTZ

Faith Builders Resource Group

Christians in the Digital Age

Faith Builders Resource Group, 28527 Guys Mills Rd., Guys Mills, PA 16327.

ISBN: 978-1-935972-34-1

Scripture quotations are from New King James Version, © 1982, unless indicated otherwise. Used by permission. All rights reserved.

Icons from Pixabay (pixabay.com). Used by permission.

Background texture on cover: Royalty-Free image from Shutterstock (shutterstock.com).

Cover Design: Courtney Brubaker

Available from Christian Learning Resource. To order or request information, please call 1-877-222-4769 or email clr@fbep.org.

Faith Builders Resource Group is a division of Faith Builders Educational Programs. Faith Builders Resource Group is dedicated to building the Kingdom of God by partnering with conservative Anabaptist communities to identify needs and to address them with services and materials that honor Christ and strengthen His church. For more information, email fbresource@fbep.org. To learn more about Faith Builders Educational Programs, visit www.fbep.org.

Printed in the United States of America.

Table of Contents

Introduction | 5

Part I: Clarifying Our Values | 9

Lesson 1 .... Love for God 11

Lesson 2 .... Love for Others 15
Our Mission

Part II: The Impact of Technology | 29

Lesson 3 .... Tools and Machines 31
Communication

Lesson 4 .... Information 37
Computer Processing

Lesson 5 .... Entertainment 43

Part III: Identifying the Traps | 47

Lesson 6 .... Gadget Lust 49
Waste of Resources

Lesson 7 .... Addiction 55
Escape

Lesson 8 .... Self Over Relationships 65

Lesson 9 .... Sexual Snares 71

Part IV: Strategies to Flourish | 77

Lesson 10 .... Assessment 79

Lesson 11 .... Connection 85

Lesson 12 .... Creativity 91
Quiet

Lesson 13 .... Freedom 99
Learning

Appendix | 109
Endnotes | 115

*Note: This material is laid out in thirteen lessons to correspond to the typical thirteen weeks in a quarter. For those settings where twelve lessons fit better (to accommodate Christmas or Easter, for example), the first two lessons could be combined to give the foundation for the remaining lessons.

Introduction

The digital age has taken the world by storm—some liken it to a tsunami. Our world is enraptured with novel gadgets and new applications, but Christians are recognizing that along with the conveniences come many hazards. Phones seem to be smarter than many people using them these days. In the past, conservative Anabaptists often drew lines to deal with dangers—no radio, no television, and so on. With the crossover of devices and technology, however, drawing lines is much more complex these days.

Recognizing that individuals and groups will draw lines at different places, this study is designed to highlight the issues and enable us to face the considerations necessary to make wise decisions. The approach is neither anti-technology nor pro. It does recognize that technology is changing our world and is intended to help us avoid simply adjusting to change without thoughtful evaluation.

An approach that is neither wholesale for nor against, however, must not assume that technology—digital technology in particular—is a neutral medium. It would be naïve to think that smart phones or the internet or PowerPoint presentations are completely non-influencing media, their effects dependent only on the heart of the user. These things are changing our cultural landscape. Using them will shape us. Smart phones are shaping language, influencing thinking, and even forming the self-concepts of the users. The internet is changing our habits, from shopping habits to relational habits to habits of evaluating ideas and people. In short, the medium plays a significant part in the outcomes.

We offer these studies, then, as a way to pause and think, as a way to look at these issues in the company of brothers and sisters in Jesus, and as a way to evaluate our hearts, our practices, and our values.

The studies are divided into four parts. Part I explores the two greatest commandments and the mission God has given us, the light by which we must evaluate all our activities and choices. The intent here is to provide a framework for the discussions that follow. Part II focuses on how technology in general, and digital technology in particular, has impacted our lives. This is an attempt to stand back and honestly consider changes that have happened and are happening. Part III explores the dangers, the ways digital technology may detract from obedience to the greatest commands and from the mission God has given us. In Part IV, we think about what strategies will help us to go forward rather than faltering.

The goal then is not to dictate to families and churches: to text or not to text, to allow PowerPoint presentations on Sunday morning or not to allow them. Rather, it is to look carefully at the practices and possibilities we encounter through digital technology so that we can make informed decisions. The reality is that in some situations a particular practice may be very helpful, and in another it might prove to be distracting. Furthermore, groups (or individuals) are sure to come out at different places, one with a view to the advantages in using a particular device and another recognizing the dangers that come with using that same device.

Paul's advice to the Roman Christians regarding kosher food helps us to see that two people may have opposite practices with a united intention: "He who eats, eats to the Lord, for he gives God thanks; and he who does not eat, to the Lord he does not eat, and gives God thanks" (Romans 14:6b).

And Paul's guidance for mutual respect is likewise helpful: "Let not him who eats despise him who does not eat, and let not him who does not eat judge him who eats; for God has received him" (v. 3).

A wrap-up at the end of the study poses the questions that families and congregations need to consider to carry this forward.

CLARIFYING OUR VALUES

—— PART I ——→

Lesson 1: Love for God

Lesson 2: Love for Others
Our Mission

Introduction to Part I: Clarifying Our Vision

In his book *Desiring the Kingdom*, James Smith writes, "Because I think that we are primarily desiring [creatures] rather than merely thinking things, I also think that what constitutes our ultimate identities—what makes us who we are, the kind of people we are—is what we love." A few pages later, he carries forward the implications of this stance: "Being a disciple of Jesus is not primarily a matter of getting the right ideas and doctrines and beliefs into your head in order to guarantee proper behavior; rather, it's a matter of being the kind of person who loves rightly—who loves God and neighbor and is oriented to the world by the primacy of that love."[1]

If he is correct—that our lives are guided more by what we love than by what we believe—then when we approach a subject like electronic technology to try to evaluate how we ought to live, we must look through the lens of our heart, not only the lens of our head. To clarify, we must not assume that our beliefs mean nothing. Both truth and lies are powerful. One leads to liberty and life, the other to bondage and death. But the reality is that what we set our hearts upon has a stronger interface with the choices we make and the habits we form than the set of beliefs we affirm.

To illustrate this point, we could ask any Christian whether God and His kingdom are more important than a new cell phone or computer, and we would likely get the correct answer. We know that God and His kingdom are more important than gadgets. This is what we believe. But will that belief actually guide our choices and direct our behaviors? The disheartening reality is that if we focus primarily on what we believe and neglect the love of our hearts, a huge gap can open between our "faith" and our actual living.

In this study, we want to think well. We want to generate discussion and come up with creative ideas about the usefulness of technology as well as its dangers. But we want to do so in the context of hearts that are set right. Without a right focus of who and what we love, these studies won't likely make much actual impact on how we follow Jesus in the digital age.

Love For God

Objectives:
- To reflect on the human capacity to love as central to being human.
- To explore what it means to love God with all our heart.
- To review expressions of love for God.

○ ○ ○ ○ ○

Love for God

"Then one of them, a lawyer, asked Him a question, testing Him, and saying, 'Teacher, which is the great commandment in the law?'

"Jesus said to him, '"You shall love the Lord your God with all your heart, with all your soul, and with all your mind." This is the first and great commandment. And the second is like it: "You shall love your neighbor as yourself." On these two commandments hang all the Law and the Prophets'" (Matthew 22:35-40).

The number one thing God wants from us is our love. He calls us to set our affection upon Him above all else, to make our devotion to Him supersede all other loyalties, and to worship Him as the highest good in time and eternity. This stance of the heart is according to how things are. God is our loving Creator. He created us to live in relationship with Him. Just as a car is engineered to drive on prepared roads across the land, not to sail on the sea, so God created us to live in loving, worshipful relationship with Him. If we choose to love ourselves or material things above God, it's like trying to take a car into the surf. However superb the technology of the car, drive it into the water and the engine will die and the car will go down.

And so, God's warning, "Do not love the world or the things in the world" (1 John 2:15), is not to keep us from enjoying life. It is to preserve us in life, to keep us from drowning. It's not the fault of "things" that they don't satisfy us. Our hearts simply were not made to thrive apart from loving commitment to God above all else. As Henry Drummond explained:

> There is a great deal in the world that is delightful and beautiful; there is a great deal in it that is great and engrossing; but it will not last. All that is in the world, the lust of the eye, the lust of the flesh, and the pride of life, are but for a little while. Love not the world therefore. Nothing that it contains is worth the life and consecration of an immortal soul. The immortal soul must give itself to something that is immortal.[2]

One of the amazing things about love is the interactivity of heart and action. Jesus taught that it is out of our hearts that we speak and act. He also told us to lay up treasure in heaven because "where your treasure is, there your heart will be also" (Matthew 6:21). So on the one hand, the heart is the source of action, and on the other hand, our actions (and particularly our habits) shape our hearts. In other words, the more love we have for God in our hearts, the more we will express that love in word and action, and the more we cultivate expressions of our love for God in words and actions, the more love we will have for Him in our hearts. The heart and its expressions have reciprocating effects on each other.

Technology has provided us with more tools than ever before in the history of mankind. We have the means of expressing and recording and recalling and transmitting expressions of love for God more than any generation before us. But love has always been and will always be a heart function. No computer will ever love God. No smart phone will gain that capacity. These are machines. They don't care if they are used in worship of God or as instruments of the worst of human sins.

Because of technological advancement, there are more "things" in the world to lust after than ever before. And human beings seem to have an exceptional fascination for electronic gadgets. These things have glitz and amazing capacities. But we must not set our hearts on them, though at times we may find them useful, even in expressing our love for God.

God designed us to love Him above all else.

Processing ·----------------->

How can we deepen our love for God?

1. Put it into words. Married couples enable their love to grow by expressing it, sometimes spontaneously, sometimes in finely crafted poetry or set to music. Our love for God likewise begs to be uttered.

    a. Make a practice of expressing your love for God out loud, spontaneously, through the day.

    b. Try writing it thoughtfully, using either prose or poetry, but choosing the best words and literary form in your ability.

    c. Read Psalm 63 aloud. Try writing it verse by verse in your own words.

    d. Find songs that express love for God and sing them to God. Examples: "Jesus, My Lord, My God, My All"; "Oh Love That Will Not Let Me Go"; "Father, I Adore You".

2. Meditate on God's love for you. Jot down the ways His love has found expression. Journal your reflections or spend time prayerfully responding to God's love.

3. Do a group activity with others who love God. Take turns finishing these sentences: "I love God because...." "I need to grow in love for God because...." "If I truly love God with all my heart, it means...."

4. Do a study in the Psalms of expressions of love for God. What do you observe about why the writer loved God or how he expressed it?

5. Choose a favorite expression of love for God (either something you have written or one you found in a hymnbook or prayer book) and read it meditatively on a scheduled basis.

6. Find a copy of your baptismal vows (or something similar if you can't find an exact copy). Read them aloud to God and meditate on what they mean for you now. You might try reading them regularly for a time.

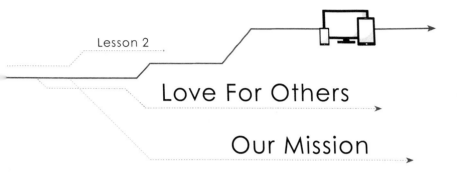

Love For Others

Our Mission

Objectives:
- To define love for others as taught and exemplified by Jesus.
- To explore specific characteristics of love in the church and the family.
- To define our mission in creation and as followers of Jesus.

○ ○ ○ ○ ○

Love for Others

In Jesus' reply to the lawyer's question, "Which is the great commandment?" He gave the man more than he asked for. "The second is like it," Jesus said. "'You shall love your neighbor as yourself'"(Matthew 22:39).

We should note a couple of things. First, we have two commands here, not one. It is true that one way we show love for God is by loving others, but loving others does not replace loving God. The one is second, the other is first. God is a person, and God is to be loved above every other person. Indeed, it is only by loving God that we are enabled to love others. But more on that later.

Some people have found God so mysterious because He is spirit, and their efforts to know and love Him is so different from the sensory love of humans, that they simply give up loving Him and focus on loving others. This we must not do, for we can make an idol of human love as easily as anything else. Nothing should replace God as the focus of our highest love, our chief delight, our unbridled worship, our supreme devotion.

But then, when we love God, we are called into this great (and impossible) task of loving others.

So the second observation is that while we cannot merge the two commands, neither can we separate them. He who loves God must love his brother. As the Apostle John put it:

> Beloved, let us love one another, for love is of God; and everyone who loves is born of God and knows God. He who does not love does not know God, for God is love. In this the love of God was manifested toward us, that God has sent His only begotten Son into the world, that we might live through Him. In this is love, not that we loved God, but that He loved us and sent His Son to be the propitiation for our sins. Beloved, if God so loved us, we also ought to love one another (1 John 4:7-11).

Loving others involves our feelings, but the love we are called to exercise is deeper than our feelings. Even in the best of relationships, feelings come and go. Love for others is a heart commitment to what is best for them. This encompasses all of life and all our relationships. In a word, love captures God's will for our lives, or as Paul put it, "Love is the fulfillment of the law" (Romans 13:10).

○ ○ ○ ○ ○

Love in the Christian Community

The night before His crucifixion, Jesus told His disciples, "A new commandment I give to you, that you love one another; as I have loved you, that you also love one another" (John 13:34). The command to love, of course, was ancient. But by His life and example, Jesus gave love new dimensions, and He called his followers to love as He loved. This, He said, would become the distinguishing mark of being His follower: "By this all will know that you are My disciples, if you have love for one another" (v. 35).

In his book *When the Church Was a Family*, Joseph Hellerman says love finds specific expression in the Christian community.[3]

1. We share our stuff with one another. "But whoever has this world's goods, and sees his brother in need, and shuts up his heart from him, how does the love of God abide in him?" (1 John 3:17).

2. We share our hearts with one another. "O Corinthians! We have spoken openly to you, our heart is wide open. You are not restricted by us, but you are restricted by your own affections. Now in return for the same (I speak as to children), you also be open" (2 Corinthians 6:11-13).

3. We stay, embrace the pain, and grow up with one another. "But, speaking the truth in love, may grow up in all things into Him who is the head—Christ— from whom the whole body, joined and knit together by what every joint supplies, according to the effective working by which every part does its share, causes growth of the body for the edifying of itself in love" (Ephesians 4:15, 16).

As followers of Jesus, we find that loving one another is not always easy. The individualism that has come with modern western culture has made it easy (even honorable) to divide, to go our own way, even to be a non-church-member Christian.[4] When the church decides to say no to particular behaviors for its members (sometimes even those specifically forbidden in the New Testament), it seems increasingly common to hear judgments like "legalism" or "Pharisaism," and for individual believers to want no part of such restrictions.

Of course, whether we are addressing the use of new technologies or older devices and activities, we must not assume that regulations will equal or always result in godliness.

But if we truly are committed to what is best for our brothers and sisters in Jesus, to doing all things in love, to doing only what edifies, to avoid causing weaker brothers or sisters to stumble—kind of commitment surely will guide (and sometimes limit) the use of new technologies. If the Apostle Paul could envision not eating certain meat for the good of his fellow believers, we wonder what he might intentionally do without in the digital age.

Processing ▸

How can Christians develop love for one another?

1. Hospitality: Christian families and singles mingling with each other in informal settings, sharing meals, playing games, discussing joys and sorrows, and learning to know each other—this should be a regular part of "loving one another fervently with a pure heart" (see 1 Peter 1:22).

2. Discipleship relationships: Older men can choose to meet with younger men, older women with younger women in one-on-one sharing times and then praying together. If you are not actively engaged in such a relationship, make a list of the persons in your life with whom you might engage or who might benefit from kind and caring interaction with you. Choose one, and write out specific steps you could take for one-on-one interaction, either as a one-time event or an ongoing practice.

3. Small groups: Christian friends benift from meeting on a scheduled basis for deep sharing. It is helpful for these groups to have clear goals. There are Bible study groups, prayer groups, accountability groups, mentoring groups, and groups formed around common experiences (such as grief, addiction, or abuse).

4. Sharing times: Often congregations have this as part of the morning worship, but this can also be done informally in after-church interaction, in youth groups, or in the family setting. Sharing is often testimonial or confessional, and includes answers to prayer, experiencing God's leading or provision, requests for prayer, acknowledgement of failure or sin, special insights God has given, etc. Many of the psalms are testimonials, and sometimes the psalmist talks about the importance of praising God and thanking Him "in the assembly."

5. Keep a list of people with special needs—elderly, hurting, or financially struggling. Bring these people to the Lord and ask Him to show you ways you might bless them.

6. Choose someone in your congregation to pray for in a focused way for a week. God may prompt you to follow it up with a note of encouragement or a different expression of support.

7. Choose one of your church leaders and do the same.

Love in the family

One of the saddest and most horrific consequences of godless individualism in modern culture is the breakdown of the family. Married couples separate and divorce. Families break apart. Broken families attempt to join one another in a variety of ways. Unwanted, unborn babies are poisoned or snipped apart in the womb and discarded. The times continue to grow more perilous as people become "lovers of their own selves" and "without natural [familial] affection" (2 Timothy 3:2, 3, KJV).

While we will explore this more later, modern technology is being used to further the breakdown of the family in a host of ways, from the medical field (as in providing abortions and abortifacient methods of birth control) to entertainment (movies and shows and online pornography sites that advocate moral deviation) to communication (texting, sexting, and social sites enabling lewd and unfaithful lifestyles).

Against this tide of familial breakdown, the call of God goes out for husbands to love their wives sacrificially, for wives to respect their husbands, for parents to turn their hearts toward their children and children toward their parents. Selfless love is the key to strong families. God's directives and prohibitions are for our good, to preserve life and happiness.

If we are to survive in an age where self has become king and where sin is ever more accessible, we must value family relationships. We must learn how to maintain and nurture the natural affection in family relationships. We must practice the activities that build and strengthen family relationships and avoid those that fray relationships. We must cultivate communication skills that nurture trust and make for deep and meaningful family ties. We must maintain family traditions that hold families together and to which families love to return. We must work to protect the childhood of our children, not expose them to adult sin. And we must learn the habits of forgiveness, and acquaint ourselves with the paths of reconciliation to heal the effects of sin within family structures.

Processing. --------------➤

What are practical ways to build joyful family relationships?

1. Doing things together—eating, playing, working, worshipping, reading, discussing plans, etc.
    a. Parents reading good stories to their children.
    b. Camping out or just sleeping out under the stars .
    c. Fishing or hunting together.
    d. Regular Bible reading.

2. Meaningful traditions—holiday traditions, special meals, story time, mushroom hunting, going for ice cream after a work evening, annual corn roast, listening to Handel's Messiah, singing together, regular attendance of church activities, and prayer together before bed time.

3. Learning together—visiting museums or sites of interest, bird watching or other specialized studies or hobbies, and doing research on an item or place of interest.

4. Compassion projects—helping a neighbor, "adopting" an elderly person or single person who doesn't have local family connections, raising funds for a special need, "adopting" a missionary, or sponsoring a child in a foreign country. Any of these kinds of projects raises awareness of the needs of others and nurtures enduring values in family life.

5. Entering each other's worlds—parents getting down on the floor with their children, children going along with parents in their responsibilities, husbands helping their wives with household duties, and wives assisting their husbands.

6. Expressing affirmation—saying thank you, pointing out good habits or thoughtful actions, and offering words of encouragement. In the family we typically know each other better than anyone outside the family; it makes sense that our best encouragement should come from each other.

Our Mission

Many years ago, I was talking to a man who lived on the streets of New York City. I do not know if the story he told me was true, but he said that he had been a successful businessman. One day he woke up and the question hit him, "What is the point of what I do?" He realized that he did not have purpose in life beyond his immediate work, and he fell into such despair that he walked away from it all.

My immediate thought was, *What a tragedy to just walk away from a successful business!* But as I pondered further, I realized this man was actually a step ahead of millions of people, some even in the church, who are still trying to find life's greatest purpose in their own pursuits. To have no purpose often leads to despair. But to have wrong purpose is likewise a tragedy. To know who we are and why we are here is to have an anchor point for life.

God has not left us without direction for who we are and the reason for our existence. A number of Scriptures clarify His intentions for us.

○ ○ ○ ○ ○

Creation mandate

Going all the way back to the beginning, we have the creation mandate:

> So God created man in His own image; in the image of God He created him; male and female He created them. Then God blessed them, and God said to them, "Be fruitful and multiply; fill the earth and subdue it; have dominion over the fish of the sea, over the birds of the air, and over every living thing that moves on the earth" (Genesis 1:27, 28).

Here we learn that we were created by God in His image and that we have been given the privilege and responsibility of caring for the earth. Of course, the human plunge into sin has badly skewed our situation. The earth now produces weeds and a host of other obstacles to comfortable living, and we have turned selfish by nature, wanting to go about things in our own way and assuming we have the wherewithal to find happiness and meaning on our own.

But even with the earth groaning under the curse, the creation mandate continues to provide a framework for our lives. Much technology has been developed to make work more efficient and to improve working conditions. Harnessing water, wind, heat, gravity, and electricity, we have made machines that cultivate the soil, make other machines, whisk us from place to place, and enable us to communicate all around the globe. Unfortunately, these inventions and "progress" have too often been fueled by greed and pride, and thus have been marred by such realities as exploitation of workers, massive accumulation of wealth, toxic waste materials, misuse of land, and terrible destruction of humankind.

The creation mandate gives us much to ponder as we consider the subject of technology.

○ ○ ○ ○

The Kingdom of God

Into the mess made by human sin, God has launched His plan of redemption through His Son. Jesus came proclaiming the kingdom of God—the restoration of life lived in obedience to God's reign. In the training of His disciples, Jesus told them not to lay up treasure on earth, not to be overly concerned even with food and clothes, but to "seek first the kingdom of God and His righteousness, and all these things shall be added to you" (Matthew 6:33). This is a call to bring all that we are and all that we have back under God's reign and to live in trusting relationship with Him. It is a radical reorientation from self-government to acknowledging God as the One who made us and knows what is best for us.

Jesus taught us that the number one thing the Father wants from us is our love, and that loving relationship with the Father will cause us to love others and serve them. He taught us that the full restoration of the kingdom of God is yet to come, that we should set our hope on this eternal kingdom, and that by living in love now, we will already be living in the ways of God's kingdom still to come. Thus we are in the "in between" era, where we are already being transformed into what we shall be fully then.

○ ○ ○ ○ ○

Making Disciples

Just before He ascended, Jesus said, "All authority has been given to Me in heaven and on earth. Go therefore and make disciples of all the nations, baptizing them in the name of the Father and of the Son and of the Holy Spirit, teaching them to observe all things that I have commanded you; and lo, I am with you always, even to the end of the age" (Matthew 28:18-20).

Here we have our primary life mission summarized. We are to proclaim Jesus and His way of living to all people, inviting them back into relationship with the Father. Those who believe in Jesus as the Way back to God are "marked out" with baptism.

Our mission is to tell the good news and enable one another to know and follow the teachings of Jesus. This is our life purpose. It gives us reason to get up every morning. It provides the rationale for work and business. It shapes how we relate to our heritage, how we order the current habits of our lives, and how we teach and train our children as we look to the future.

The ways of Jesus, of course, are as broad as life. We can eat and drink, work and play, worship and laugh, invent and rest, all for the glory of God. The mission we have been given calls us in all that we do to consider the overarching purpose of our existence, and specifically the will of God for each person and how he or she will contribute to the mission. We are called to bring all the aspects of our lives to serve His good purposes.

The followers of Jesus comprise His body. Just as when Jesus was on earth, He "went about doing good" (Acts

10:38), so His followers now do the same. We are to have the same kind of values, the same care of people, the same love of righteousness, the same delight in the Father, and the same willingness to use our resources for the good of others.

Each follower of Jesus, of course, contributes in his own way, even as every member of our physical bodies makes its unique contribution to the whole. Among God's people, some lead, some teach, some give, some help, some help the helpers, but all contribute in some way to carrying out the will of God. While certain kinds of activities contribute more directly to the spread of the Gospel or the care of souls, every member's contribution is important. Just as in an army, front-line soldiers could not succeed without the behind-the-scenes work of engineers and cooks, so in the church, missionaries and pastors need the support of those who pray and give and do accounting.

The mission to invite others to join God's people and enable them to follow Jesus provides us with a framework for thinking about technology—how to utilize it, what is worth our time and attention, and what is not.

Processing......................➤

1. As you consider western Christians, make a list of what you think are the biggest hindrances to Christians fulfilling their mission.

2. As you consider conservative Anabaptists, what would you add/take from this list?

3. As you consider your own life, what would you single out from your lists as your biggest hindrance?

4. In what ways do you see modern technology fulfilling the creation mandate, and in what ways is it disregarding that mandate?

5. Personal assessment:

   a. Make a list of your skills.

   b. What would you say is your primary spiritual gift? (You can also list additional gifts.)

   c. What have you done to develop your abilities and gifts, or what plans do you have for developing your gifts?

   d. What material things/resources do you have that might contribute to the kingdom of God? Alternate question: In what way(s) have you squandered material things that might have contributed to God's kingdom?

   e. As you think about all that God has given you, describe as clearly as you can how you see yourself contributing to the mission of God's people.

   f. As you think about God's will and your life, list any changes you would like to make.

THE IMPACT OF TECHNOLOGY

●── PART II ──➤

Lesson 3: Tools & Machines
        Communication

Lesson 4: Information
        Computer Processing

Lesson 5: Entertainment

Introduction to Part II: The Impact of Technology

In 1970, Alvin Toffler published his book *Future Shock* in which he described the "accelerative thrust" of change, fueled largely, he said, by technology. Using transportation as an example, he noted that for many centuries,

> ...the fastest transportation available to man over long distances was the camel caravan, averaging eight miles per hour. It was not until about 1600 B.C. when the chariot was invented that the maximum speed was raised to roughly twenty miles per hour.
>
> It was probably not until the 1880's that man, with the help of a more advanced steam locomotive, managed to reach a speed of one hundred mph...
>
> It took only fifty-eight years, however, to quadruple the limit, so that by 1938 airborne man was cracking the 400-mph line. It took a mere twenty-year flick of time to double the limit again. And by the 1960's rocket planes approached speeds of 4,000 mph, and men in space capsules were circling the earth at 18,000 mph. Plotted on a graph, the line representing progress in the past generation would leap vertically off the page.
>
> Whether we examine distances traveled, altitudes reached, minerals mined, or explosive power harnessed, the same accelerative trend is obvious. The pattern, here and in a thousand other statistical series, is absolutely clear and unmistakable. Millennia or centuries go by, and then, in our own times, a sudden bursting of the limits, a fantastic spurt forward.
>
> The reason for this is that technology feeds on itself. Technology makes more technology possible."[1]

When Toffler wrote this, there were no cell phones, no laptop computers, no smartphones, no CDs, no mp3 players, no GPS devices, no drones, no computerized carburetors, no high-definition TVs or projectors, no endoscopic surgeries, and no social media. There was no internet. And incidentally, as I write this, the spacecraft Juno has entered Jupiter's orbit, having slowed down from its traveling speed of 165,000 mph.

Our world is changing.

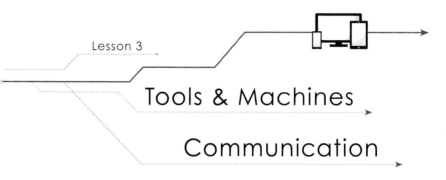

Lesson 3

Tools & Machines

Communication

Objectives:
- To reflect on how technology has produced tools and machines to make our work easier.
- To evaluate how the changing means of communication has affected the quality of our communication.

○ ○ ○ ○ ○

Tools and Machines

Manual labor is fast disappearing. We employ labor-saving devices in virtually every area of our lives, and we do less and less "by hand." Machines bake our meals and wash our dishes. Machines sew our clothes and clean our floors. Machines work our ground and harvest our crops. Machines cut and shape and fasten our wood to make houses and cabinets and furniture.

Tools have changed our lives, and new tools are constantly being produced and marketed, sometimes making older tools obsolete. Carpenters rarely drive nails with a hammer, for example, or use a manual screwdriver. Nail guns and battery-powered drivers are faster and take less effort. We still use pencils and pens, but a typewriter is a machine of the past. It went out with the record player and the eight track.

Tools make our work more efficient and our lives more comfortable. We can build things faster and with less waste. A cabinet shop can use a computerized laser cutter that calculates how to cut the necessary parts out of a sheet of plywood with the least waste. A computer program can design a truss to meet load specifications, again with the most efficient use of materials. On occasion a man might use a shovel to dig, but for most jobs, it is a far better use of his time to get a mechanical digger to do the work.

Processing

1. Make a list of tools and machines you have purchased in the last year. Separate the replacement tools from tools that are new for you.

2. For each tool or machine, think about how your life would be different if you did not have it.

3. Reflecting on how you used to do things, write a summary of how tools and machines have changed your life.

4. As you look over your list, are there any tools or machines that have not worked out as you thought they would? (Maybe they aren't as useful as you thought they would be, or you use them less than you thought you would, or they created other "needs" that you hadn't expected, etc.)

5. Are there times you have intentionally not purchased new tools or machines? If so, try to articulate your reasons.

Communication

In the year 2000, I spent several months in Romania. A missionary friend of mine told me that when he first served overseas in the 1960s, he was gone from home for two years and had only the occasional letter exchange and no phone contact with his family during that two-year stint. He lamented that a group of young people had recently visited the same mission from the States, and he was ashamed for his Romanian accountant to see the phone bill for that month, even though the youth had left money to pay for their use of the phone. Today, of course, they could use their smart phones with an international plan, bypass the long-distance charges, and not only talk, but video chat with family and friends.

Technology has certainly changed communication. We can hardly imagine life without email or cell phones. Phones enable us to stay in communication either by voice or by text with nearly anyone at nearly any time of day or night. We can talk, text, Snapchat, Skype, or email. If we want to communicate something visual, we can scan it, fax it, or take a picture and send it. Even as I write this, I am planning to go on an outing with some students. I pulled up a detailed map of the area where we are going and snapped a picture so that I can retrieve it when we get close to our destination. Or I could punch in the address (or coordinates) in my GPS and follow that, complete with vocal directions.

Technology not only changes the means by which we communicate, but it also changes our communication in other ways. We fire off quick questions or comments and get quick replies. We grow impatient if we can't contact someone or if we don't hear back immediately. The quick and efficient mentality often shaves off unnecessary letters and words, especially in the texting world.

"Ppl plz turn down ur music b4 u go deaf! Lol!"

"K thx"

Of course new technology introduces new vocabulary, too. We hear words like selfie, iPhone, smartphone, unfriend, texting, emoji, techie... it can be difficult to keep up with the "in" words and terms, especially if one doesn't use the latest communication technology.

While technology has enabled us to connect with each other in unprecedented ways—quickly and across long distances—electronic communication also has its traps. We will discuss these in more detail later, but words transmitted via text are often written in haste and lack the visual and aural clues that supply the emotional dimensions of communication. A text that says "K" for okay, doesn't tell us if the person is happy, sad, disappointed, angry, or excited. This easily results in misreading responses in texts and emails. Furthermore, communicating from a distance provides a layer of anonymity for those whose intentions are not good.

Processing.----------------➤

1. How has technology changed your communication in the past two years?

2. Can you give examples of miscommunication you have experienced due to haste or lack of presence?

3. What kinds of communication do you think especially require presence?

4. What are the advantages and disadvantages of instant communication?

5. Initiate a conversation with someone close to you to evaluate your communication. Ask the person to describe honestly how technology has changed the way you communicate. How do you see these changes in communication affecting your relationship—positively and negatively? Are there communication habits from earlier in your relationship that have been discontinued or replaced? Are there communication habits you enjoy that have been made possible through technology? If you could change something about your spouse's (or friend's) communication via technology, what would it be?

6. What is the most annoying habit/action you observe in other people's use of cell phones?

7. What courtesy rules or protocol would you like to see established for cell phone use?

8. What considerations should guide parents in determining when it is appropriate for their children to have a cell phone?

9. What guidelines for cell phone use are helpful for first-time users?

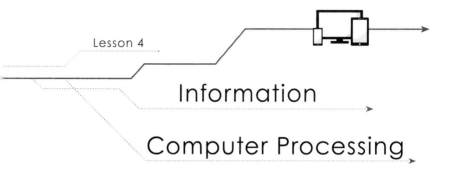

Lesson 4

Information

Computer Processing

Ojectives:
- To evaluate the impact of access to information.
- To reflect on the way computer processing is changing our lives.

○ ○ ○ ○ ○

Information

The Internet has rocketed us into the Information Age. Instead of reading the newspaper at the end of the day, we get real-time alerts of headline news or bad weather on phones and computers, giving us the opportunity to explore further if we wish. Instead of going to the library to do research, we can do a search and almost instantly have a trove of articles, blogs, and entire websites devoted to the subject. If it is a controversial subject, we will get extreme pros, extreme cons, and all the shades in between.

All this available information impacts us in many ways.

The sheer amount of information can create overload. There can be dozens of news articles about the same incident. On controversial issues, the information is bewildering. How can a person process all the voices shouting every possible angle of an argument, each claiming to be the highest authority?

Another result of the Information Age has been what analysts call a "leveling" of authority. A patient goes to a doctor and receives a diagnosis and a prescribed regimen for recovery. The patient goes home, does his own research, and decides to take an alternate route, or a combination of what the doctor said and alternative options he discovers online. The leveling simply means that experts are no longer in the elevated position of knowing best.

In education, the result has moved beyond providing content to educating students how to access and evaluate informantion. But of course, the very efforts at teaching others how to evaluate can come into question by the "superior" evaluation of online and offline authorities. A person who has a bent toward knowing more than anyone else can find all sorts of backing for "what really happened" in the latest election, national tragedy, or public scandal.

The practical uses of technological information are almost unlimited. If a person wishes to bake a perfect loaf of whole wheat bread or change the alternator on a car, all he or she needs to do is to look it up on the internet. There will likely be multiple video clips showing step-by-step how to do almost any task.

Processing.----------------►

1. What is the most common way you use technology to access information?

2. How have you seen positive and negative effects of the leveling of "authority"? Have you observed people using and trusting online sources in unhealthy ways?

3. What criteria do you use to evaluate sources of information?

4. How has the availability of information affected your work?

5. Do you find yourself surfing for information, using your phone or computer to browse? If so, how have you seen this affecting you, your use of time, or your relationships?

Computer Processing

We call them computers. They started out adding, subtracting, multiplying, and dividing. But now they sort, categorize, analyze, store, retrieve, chart, evaluate, predict, and even talk to us.

The processing speed of computers has simply leapt off the charts. We now hold cell phones with processors faster than personal computers were only a few years ago. In 2016, a team at the University of California produced a microchip that had 1,000 independent programmable processors. According to a report in Science News, "The energy-efficient 'KiloCore' chip has a maximum computation rate of 1.78 trillion instructions per second and contains 621 million transistors."[2] Bavan Baas, professor of electrical and computer engineering, led the team in producing the microchip. He says, "The KiloCore chip executes instructions more than 100 times more efficiently than a modern laptop processor."[3]

It is difficult to stay current with the way computer processing has changed and is changing our lives. Credit cards and debit cards, of course, have been with us for some years, but we now have on-line banking where we can keep track of transactions as they occur, and we can deposit checks by snapping a picture with our cell phones. Technology enables businesses to keep track of inventory, fill orders, identify what sells and what doesn't, calculate profit and loss, and keep supplies stocked automatically. Technology enables scientists to organize and analyze data and then predict outcomes. Technology enables farmers to put chips on their cows that keep track of their productivity, and feed them accordingly.

Computers can process information with speed and efficiency far beyond human ability—which of course, raises the fascinating (horrifying?) possibility of computers becoming the masters of humanity, a subject the sci-fi world regularly imagines.

In 1996, IBM programmed a computer named Deep Bule to play chess at a master's level. Garry Kasparov, the reigning world chess champion, was able to beat the computer 4-2. The next year, Deep Blue came back to beat Kasparov 3 ½ to 2 ½. In a fascinating article published in Time online, David Gelernter commented on the difference between a machine and a human being. He asks the question: "Shouldn't we conclude that Deep Blue must be a thinking computer, and a smart one at that, maybe brilliant?"

He continues, answering his own question:

> No. Deep Blue is just a machine. It doesn't have a mind any more than a flowerpot has a mind. Deep Blue is a beautiful and amazing technological achievement. It is

an intellectual milestone, and its chief meaning is this: that human beings are champion machine builders. All sorts of activities that we thought could be done only by minds can in fact be done by machines too, if the machine builders are smart enough...

[W]hen you think about it carefully, the idea that Deep Blue has a mind is absurd. How can an object that wants nothing, fears nothing, enjoys nothing, needs nothing and cares about nothing have a mind? It can win at chess, but not because it wants to. It isn't happy when it wins or sad when it loses. What... if it beats Kasparov? Is it hoping to take Deep Pink out for a night on the town? It doesn't care about chess or anything else. It plays the game for the same reason a calculator adds or a toaster toasts: because it is a machine designed for that purpose.[4]

Near the conclusion of his article, Gelernter writes:

In the long run I doubt if there is any kind of human behavior computers can't fake, any kind of performance they can't put on. It is conceivable that one day, computers will be better than humans at nearly everything. I can imagine that a person might someday have a computer for a best friend. That will be sad—like having a dog for your best friend but even sadder.[5]

Although processors can never be human, they certainly have changed human life. And we do well to think about those changes carefully.

Processing.------------➤

1. In what ways have computers made your line of work more efficient?

2. Give examples of ways you have used computers to calculate, store, retrieve, or organize information in the past month.

3. How did you used to do the same things? (Or how did your parents do these activities a generation ago?)

4. How do you envision the efficiency of computers affecting human abilities?

5. Armed with computers, we can compete with other (in business, in education, in purchasing, in recreation). How might this give us a false sense of power? And can you give examples from your experience?

Entertainment

Ojectives:
- To reflect on the ways technology has shaped the means of entertainment.
- To evaluate the role of entertainment personally, relationally, and spiritually.

○ ○ ○ ○ ○

Entertainment

In 1953, George Gallup gave a speech at the dedication of a new communication center at the University of Iowa. In addressing a number of concerns about the American educational system, he made some incisive comments about entertainment: "One of the greatest threats to the freedom of the world is the citizen who wants to be entertained rather than informed," Gallup said. A bit later he continued, "This country can go so heavily toward entertainment that we may literally 'kill ourselves laughing.'"[6] We wonder what he would say about how entertainment has shifted in the intervening decades.

A statistical study done by Mike Masnick in 2012 illustrates the increasing impact of entertainment in the United States. He says:

- In 2008, 56 million Americans were playing video games; in 2011—three years later—there were 135 million.

- In 2005, there were 1,000 gaming companies; in only two years, that rose to 18,000.

- From 2002 to 2010, the music artists' share of profits for their music rose 16% to an annual income of $16.7 billion.

- In 2007, every minute there were eight hours of video uploaded to YouTube; by 2011, that had increased to 48 hours uploaded every minute.

- In 2002, there were 693 million music purchases; in 2010, that increased to 1,507,000,000.[7]

Who can calculate the total impact of video games, movies, music, live performance, and online entertainment on US culture? If high doses of entertainment dull intelligent thinking (which was George Gallup's concern), what numbing up and dumbing down must be happening in our world today? And if Gallup was concerned about the effects on serious thinking, how much more ought we to be concerned about the spiritual effects?

The Apostle Paul warned of the perilous "last days" in which people would be "lovers of pleasure rather than lovers of God" (2 Timothy 3:4).

Obviously, it is not wrong to enjoy life, to relax, to play a game, and to experience pleasure. In the same letter where Paul told Timothy that "she who lives in pleasure is dead while she lives" (1 Timothy 5:6), he spoke appreciatively of "the living God, who gives us richly all things to enjoy" (1 Timothy 6:17).

It is one thing to love God with all our heart and enjoy His good gifts to us; it is another to love the things of the world. It is one thing to find refreshment and relaxation from meaningful work; it is another to work in order to seek pleasure, to make pleasure the "end" or the measure of a good life.

Because of technology, we have unprecedented possibilities for entertainment. With recorded music, we have access to musicians, orchestras, and choirs any time we wish. With cell phones and connectivity, we can play games with people halfway around the world or watch sporting events, again nearly anywhere in the world.

And American wealth means all this entertainment is at our fingertips, only a few clicks away, day or night.

We will think more evaluatively of the advantages and disadvantages of technology in a later study. Here we are focusing more on the ways technology has affected our lives.

Processing.----------->

1. What forms of entertainment do you commonly engage in?

2. For each of these forms of entertainment, tell how you access it (technological device), or tell how technology has enabled or shaped this means of entertainment.

3. How do you think entertainment has shaped you and your friends?

4. In what ways has your means of entertainment changed in the last five years?

5. In what ways has the content of your entertainment changed over the last five years?

6. What cautions or guidelines do you observe in engaging with entertainment?

7. In what ways is the technologically provided entertainment of your children (or your parents) different from your own?

8. What are the spiritual effects of a heavy pursuit of entertainment?

IDENTIFYING THE TRAPS

⟶ PART III ⟶

Lesson 6: Gadget Lust
        Waste of Resources

Lesson 7: Addiction
        Escape

Lesson 8: Self Over Relationships

Lesson 9: Sexual Snares

Introduction to Part III: Identifying the Traps

Technology has been changing our lives rapidly and profoundly. Everything from eating habits, to the way we do business, to transportation, to recreation, to medical treatment, to communication—technology has reshaped virtually every area of our lives. The benefits from these changes are immense. We can buy fresh fruit and vegetables year-round. We have information literally at our fingertips. We can travel distances in ten minutes that once would have taken hours.

With all good changes, however, there are potential dangers.

In this section, we will consider seven common traps surrounding technology. For some people, the dangers are so high that they have chosen simply to withdraw from the use of technology (though in western society, complete withdrawal is nearly impossible). Or a similar approach is to verbally bash any new device and those who use it, even while benefitting significantly from technological advancement. The approach here is not to say technology is bad, but to identify its dangers.

In the next section, we will carry forward the challenge of how to respond wisely to the challenges, dangers, and opportunities afforded by technology.

Lesson 6

Gadget Lust

Waste of Resources

Objectives:
- To increase awareness of the lure of new elctronic gadgets.
- To evaluate our stewardship of money, time, and abilities.

○ ○ ○ ○ ○

Trap 1: Gadget Lust

When we mix the genius of invention with the potential of technological production, and then add the attraction of high-glitz advertising, we experience a virtual flood of new gadgets. And we don't even have to go shopping for them. We can access online markets from the comfort of a recliner, order what we want, pay, and watch for it to arrive on our doorstep, sometimes within hours.

Gadgets come in all shapes and sizes and colors. We can wear them, carry them in our pockets, attach them to the dashboards of our vehicles, place them on our workspace, fidget with them, hang them on our ears, talk to them, listen to them, watch them, play with them, poke their buttons or screens, take them with us to bed, and instruct them to monitor our activities asleep or awake.

All of these gadgets beg to be bought. They appeal to convenience, efficiency, excitement, relaxation, status, visual effect, gratification, individulality, health, and even spirituality.

We see, we want, we buy.

The new one today will be the outdated one tomorrow.

Where is the line between necessity and convenience?

Where is the line between convenience and luxury?

Where is the line between practicality and obsession?

At what point do we begin to love the things of this world?

"Love not!" John writes. "For all that is in the world—the lust of the flesh, the lust of the eyes, and the pride of life—is not of the Father but is of the world. And the world is passing away, and the lust of it: but he who does the will of God abides forever" (1 John 2:15-17).

Gadget lust drives us to want what we do not need, buy simply to have, lust for something new, demand the latest version, and live in that vicious round of seeing, wanting, buying, experimenting, getting excited, growing bored, and finding the next item.

Processing ------------➤

1. What are the signs of gadget lust?

2. Can you relate experiences of wanting to buy something, buying it, and later realizing that it was pointless?

3. What steps have you taken to resist gadget lust?

Identifying the Traps ──→

Trap 2: Waste of Resources

Western civilization sports the wealthiest cultures in the history of the world. What is termed the "poverty level" in the United States is far above the norm for much of the rest of the world. For example, in 2015 the average annual income per capita in the United States was $55,980, with $24,250 designated as "poverty level" for a family of four. The average annual income per capita in Canada was $47,540; in Mexico it was $9,710; in Egypt it was $3,340; and in the Democratic Republic of the Congo was $410; Incidentally, the United States placed tenth in per capita income in 2015, with Australia and eight European countries ranking higher.

We are wealthy.

And how do people in the United States use their resources? Here are a few statistics on US expenditures:

- The United States has an estimated 265.9 million smartphone users who average $50/month on phone plans, amounting to an estimated $159.5 billion annually.[1]

- Every second, $3,075.64 is being spent on pornography in the United States, and every thirty-nine minutes, a new pornography video is being created.[2]

- Americans bought an estimated 300 million "wearable devices" (including 30 million smart watches) in 2017.[3]

- In 2015, consumers spent $23.5 billion on the electronic gaming industry. [4]

Americans use of time is also troubling.

- Smart phone users spend about five hours each day on their phones.

- Children ages two to five spend more than 32 hours per week watching TV, and children ages six to eight, who spend a good chunk of the day in school, still average 28 hours per week in front of the screen.[5]

According to a Kaiser Foundation study conducted in 2010, children between the ages of eight and eighteen spend an average of 7 hours 38 minutes a day with digital media. When the use of more than one digital device at a time is taken into account, they spend more than 10-1/2 hours a day with digital technologies. Television is the largest culprit, with kids watching an average of 4 hours 29 minutes a day. Music is the second leading technology, with an average of 2 hours 31 minutes a day. Kids spend about 1-1/2 hours a day using computers and 1 hour 13 minutes a day playing video games.[6] According to a 2017 article in Time, in the years since that study, the overall time children spend on electronic media has stayed about the same, with the primary shift being toward more time using a mobile device.[7]

While traditional games like chess and word games can be played on electronic devices (either against the device or against online opponents), the most popular electronic games are games of "action" and simulated professional sports. In 2018, for example, the 4 top selling games were "Call of Duty: Black Ops IIII," "Red Dead Redemption II," "NBA 2K19," and "Madden NFL 19."[8]

Interestingly, the average age of gamers spending the most time and money is 37 years old. Of those gamers, 59% are male and 41% female. Male gaming beyond the teen years is such an obsession that we now have the term "gaming widow" to describe the woman whose husband spends the majority of his free time at a console, to the neglect of his wife and family.

Processing ·····················➤

1. Do some research: what recommendations or warnings are being given for watching (or playing on) electronic devices for children, adolescents, and adults?

2. Do some self-evaluation: how much money have you spent on electronic equipment, programs, and upgrades in the past year? What percentage of that was for business (or usefulness), and what percentage was for pleasure?

3. In what ways have you caught yourself wasting resources of time, money, or relationships in the use of technology?

Lesson 7

Addiction

Escape

Objectives:
- To explore the signs of addiction.
- To evaluate how digital technology can be addicting.
- To differentiate between healthy and unhealthy escape.

○ ○ ○ ○ ○

Trap 3: Addiction

In an article entitled "Who's Really Addicting You to Technology," Nir Eyal first zeroed in on the producers of tech devices. He wrote:

> Online services like Facebook, YouTube, Twitter, Instagram, Buzzfeed and the like, are called out as masters of manipulation — making products so good, people can't stop using them....

> Since these services rely on advertising revenue, the more frequently you use them, the more money they make. It's no wonder these companies employ teams of people focused on engineering their services to be as engaging as possible. These products aren't habit-forming by chance; it's by design. They have an incentive to keep us hooked.[9]

While this doesn't nix our personal responsibility for how we use electronic devices, it does help us to think about the addictive nature of electronic devices. Manufacturers intentionally design them to pull us into finding more and more stimulation and to pull us back when we leave.

Addiction finds expression in a variety of ways. Some people are not able to stay present in a conversation, especially in a group setting, without pulling out their phones to check for email or answer texts from friends. Others flit from one news site to the next, "surfing" for the titillating, the odd, or the exotic story. And social media saps hours every day from young people as well as older ones. In a report in the Los Angeles Times, for example, a 17-year-old girl said, "The average day for me, if I'm not at work, I will text all day or be on social media. That's my life." She went on to say that on the previous Sunday she had "fiddled around online from 9 p.m. to 2 a.m.," updating her status and commenting on her friends' pages.[10]

Although gaming addiction is generally a waste of human resources, in extreme cases gamers have actually died from exhaustion while on a gaming binge. On January 8, 2015, a man named Hsieh was found dead in a Taiwanese internet café after three days of steady gaming. Ironically, "Police said gamers in the café continued as if nothing happened even when the police and paramedics arrived." Hsieh was pronounced dead of cardiac arrest. According to the news report, he was the second online gamer to die that year in Taiwan. A 38-year-old man had died on January 1, 2015, after five straight days of gaming.[11]

In 2011, Alyse Baddley from Utah posted an ad on Craigslist to sell her game-obsessed husband. The ad read:

> I am selling my 22-year-old husband. He enjoys eating and playing video games all day. Easy to maintain, just feed and water every 3-5 hours. You must have Internet and space for gaming. Got tired of waiting so free to good home. If acceptable replacement is offered will trade.[12]

Although Mrs. Baddley insisted she posted the ad as a joke, she says she posted it after he went on a 48-hour gaming binge, leaving her feeling lonely and ignored. She received "an avalanche" of replies including a number of personal ads from men offering to trade places with her husband.

There are differences of opinion about whether we ought to call such behaviors addictive. We readily recognize, for example, that there are physiological factors involved in addiction to alcohol, tobacco, or other drugs. Can a person actually be addicted, however, to an internet game or to pornography? It may depend on what we mean by "addiction." In the English language, the word certainly has a broad usage that includes repetitive behaviors. In the KJV, for example, the household of Stephanas was commended for having "addicted themselves to the ministry of the saints" (1 Corinthians 16:15). Although Peter uses different wording, he expresses the same concept to describe bad behavior when he writes of wicked people who "cannot cease from sin" (2 Peter 2:14).

The more common scriptural language is to describe addictive behaviors, especially in the negative sense, as "bondage." This is one of the dangerous characteristics of sin. Once leads to twice. Twice leads to a habit. And when sinful habits have taken hold, they don't want to let go. Interestingly, researchers have found that the brain activity in a pornography addict is virtually indistinguishable from the brain activity of a drug addict.[13]

Whatever terminology we use, one of the dangers associated with electronic technology is the danger of returning to it again and again, not because the technology is serving us but because we have become its slaves.

What are the signs of addiction? Numerous sources have identified ten common signs of substance abuse. In the chart to the right, these ten are listed in the left-hand column. The right-hand column applies this to evaluating one's addiction to technology.

Processing ⤏

- Consider the evaluative questions in the table on the next page.

Ten Signs of Substance Addiction[14]	Are You Addicted to Technology?
• Craving—intense inner urge or need for the drug	Do you constantly have the urge to check your phone, pull up social media sites, or surf the web?
• Neglecting responsibilities—choosing a fix over meeting work or personal obligations	Does your use of media distract you at work or cause you to be irritable when someone wants your attention?
• Drug-seeking behaviors—spending time or arranging schedules to find the drug	Does your mind constantly go to when you can return to a game or connect with your friends on social media?
• Physical dependence—not only do brain and body chemistry change, but the person functions in sub-optimal modes when the drug is not in the system	Do you find yourself tuned out in face to face conversations and interactions with your mind going back to an online game or activity?
• Unhealthy relationships—friendships are formed and maintained around drug usage	How has your use of tech formed your relationships? Are you drawn to those who play the same game or spend time on the same site?
• Isolating—withdrawing from family and former friends to practice the drug use	Has your use of tech alienated you from your family or church relationships?
• Poor money management—spending resources on drugs that drain accounts or wreck budgets	Do your expenditures for programs, equipment, or site access comprise a significant amount of your income? Does it keep you from meeting other obligations?
• Poor judgment—resorting to risky or wrong behaviors such as stealing, lying, or cheating to hide or maintain drug use	Have you compromised your character or your ethics to enable your use of technology?
• Increased tolerance—needing more and more to satisfy	Do you find your desire for the latest technology increasing? Are you frequently buying the newest or latest even while the older is serviceable?
• Withdrawal symptoms—being nervous, irritated, headachy, etc. when attempting to stop	Are you able to lay aside any use of technology for extended times of quiet and reflection?

Trap 4: Escape

As we have noted before, our world is changing rapidly. Consequently, many people have become disillusioned with life for a variety of reasons. Some are disillusioned with technological advances and try to withdraw to a more simple, back-to-earth way of living. Others are disillusioned with modern science and turn to spiritualism in one form or another, such as eastern religions or Native American spirituality. Many others have been hurt in the disintegration of family life and the relational dysfunction that attends the idolization of self. For them, escape from the real world provides welcome relief.

And technology gives oodles of opportunities for such escape. So while the games and social media can seem to be merely time wasters, for many people they offer an escape from a world of pain and disappointment.

On the far side of escape is the world of virtual reality. Some sites offer virtual lives replete with virtual jobs, virtual houses with virtual furniture and services, virtual children, and virtual pets. And the amazing reality is that people will spend real money for these virtual things.

Perhaps the saddest feature of the virtual world is virtual friendship. In Japan (a society known to be on the cutting edge of technology), virtual girlfriends and boyfriends are so sophisticated that Japanese young people actually prefer virtual friendships over real ones. "There is no friction in these relationships, obviously," says Loulou d'Aki, a Swedish photographer who documented a number of Japanese participants. "The girls behave very sweetly with the guys in what they say, how they respond to them, and with big eyes and heart-shaped faces—who wouldn't want that?"[15]

A heartbreaking story emerged in the Boston Globe in March of 2010. The three-month-old baby of a gaming couple in South Korea died of starvation. According to the article, the parents fed their baby once a day while spending up to twelve hours a day playing on the internet. Ironically, the game they were playing was a virtual world that included a virtual baby.

> [They] immersed themselves in a role-playing game called Prius Online, where they were "raising" a perfect little girl named Anima…. After one such 12-hour shift… the couple came home to find their baby dead and called the police. An autopsy determined the cause of death to be prolonged malnutrition….
>
> A police officer told reporters, "The couple seemed to have lost their will to live a normal life because they didn't have jobs and gave birth to a premature baby. They indulged themselves in the online game of raising a virtual character so as to escape from reality, which led to the death of their real baby."[16]

Although these are extreme examples, escape can be a problem for anyone, especially when digital images and activities are more emotive, exciting, or pleasurable than real life.

When life is painful, when life is disappointing, even when life is boring, we become vulnerable to seeking an escape. Obviously, there are many options for escape besides the digital world. We can turn even good activities into escape, including reading, listening to music, working, and Christian ministry. What makes digital technology such an easy escape route is not only its glitz but also its ease of access. If we have a computer or a smart phone, with a few taps or clicks or voice commands, we can leave what we are doing and enter the digital world.

Using electronic media as an escape can have many consequences. It can cut productivity and efficiency if we escape while at work. It can intrude on relationships if we are tuned out to conversations and interactions with friends and family members. It can eat up huge amounts of time and resources. But perhaps most significantly, when we escape, we are not engaging with life. This means we are not processing the events of life. We are not growing. We are not learning and deepening from difficulties and problems. And consequently, we are missing the life we were created for.

We should note here that not all escape is bad. When life is heavy, when we are overwhelmed, when troubles pile up, when we are in deep grief, escape can offer relief. This may come in the form of humor, a walk, or a game. It may be as short as a few minutes or as long as a sabbatical. Healthy escape is doing something to relieve and refresh body and mind so that we can again face our responsibilities in the real world. Healthy escape is temporary respite from our normal mode of engagement with life.

In discussing the experience of grief, Mildred Tengbom says, "We who grieve and those who support us should understand that we need some times of escape as a necessary part of the grieving process.... We hurt too much. Denial and escape provide us short resting places, time to regain our courage and strength so we can go on."[17]

Escape quickly becomes unhealthy, however, when it becomes our preferred mode of living, disengaging us from our friends and responsibilities. It is unhealthy when it dulls the mind, when it leaves us lethargic and unmotivated, and when it dims our purpose for living in the real world.

Processing⟶

1. Identify a number of healthy escapes you use and enjoy.

2. What would be the signs of these escapes becoming unhealthy for you?

3. As you think about your friends and their use of digital technology, do you have concern about their useage serving as an escape? If so, try to write out your concern.

4. After evaluating, ask yourself if you use the same standards to evaluate yourself as you use to evaluate your friends.

5. What are some safeguards that might keep us from turning to digital technology as a way of escape?

Self Over Relationships

Objectives:
- To explore how digital technology has fed a narcisistic approach to life.
- To evaluate the impact of digital technology on relationships.

○ ○ ○ ○ ○

Trap 5: Self Over Relationships

One of the more subtle traps that surrounds personal electronic devices is the trend toward self-absorption and the subsequent deterioration of relationships.

Of course, the problem of self-absorption is larger than misuse of technology. We live in a culture that has pampered and petted and lotioned itself back into psychological babyhood. We are, unfortunately, loving ourselves to death—overrating our abilities, over-valuing our opinions, and admiring our selfies at every turn.

While technology may not be the main culprit, it certainly is helping along the illusions of our self-importance.

The information highway offers us the illusion of wisdom. We can access factual information about virtually any subject, making us wise in our own eyes and unwilling to be instructed by others, even the experts.

The blogosphere has helped to level the writer's playing field. Anyone with enough passion can add his or her voice to the chorus on the wide, wide web. This is not to say that every voice will be heard or that all bloggers are self-promoting. But it is to say that we have increasing outlets for self-expression, and if we have nothing substantial to say, we can nonetheless post comments about what others are saying. Unfortunately, much gets posted that never would merit getting printed. But posting feeds the illusion of self-importance.

And of course, the ease of taking pictures and instantly seeing the results (and posting them for others to see) has turned every event into an opportunity to take a selfie… beside a monument, with a friend, at an event, at a celebration, feeling goofy, or just because we like to see ourselves and show ourselves to others.

We have become absorbed in ourselves, and our gadgets help us along.

In addition, these gadgets have significant appeal themselves. Watch people walking down a sidewalk or in a store. Observe how many of them are holding a phone in front of them as they walk—checking email, texting a friend, looking up something online, playing an online game. Watch families sitting in a restaurant, or even a group of friends, and observe how many are looking at a phone even as they visit, or are wearing earbuds and are oblivious to those around them. Unfortunately, the very gadgets that enable us to connect to anyone at any time also keep us from connecting with those who are present.

Even more subtle is the illusion of intimacy via electronic media. The internet offers discussion forums and dating sites for people to interact, post personal information, and look for prospective partners. Texting provides opportunity to chat with friends both legitimate and illicit. The upside of instant communication, of course, is the ability to touch base frequently, make last-minute changes to plans, and connect across the miles. The benefits are especially helpful for existing relationships, where we have significant interaction besides the texts and emails. Those benefits disappear quickly, however, when we rely on electronic communication to learn to know someone. Texts, selfies, and videos can give the illusion of deep and personal interaction even while being very selective and superficial in actual acquaintance.

In an ironic (and tragic) symbol of our self-centered and relationally defunct times, a lady who was unable to find "the man of her dreams" before turning forty, decided to marry herself. On January 27, 2015, Yasmin Eleby "was walked down the aisle by her mother and 'married' by three ministers—one of whom was her sister. As one cannot legally marry one's self in America, the ceremony was spiritual rather than legal." Eleby's bridal party consisted of ten bridesmaids. She said after the wedding that she "couldn't imagine the ceremony being any more poignant and meaningful." She added, "I was overwhelmed with the outpouring of love and support that was shown to me during my celebration of love and life."[18]

Thinking along similar lines, Katie Medlock encouraged people to be their own valentine for Valentine's Day, 2016.

It's that time of year again, folks. When we turn to the one we love and shower them with affection, warmth and love. When we take time to plan something they will really enjoy. When we reflect on what they need to feel cherished and pampered. If this all sounds like it would be wonderful to receive, as well, why not be the one to give it to yourself?

Even if you already have a valentine this year, also try out being your own. There is no other person with whom you spend every single second of your life and take care of around the clock—and if it feels as if you do this for other people in your life, then you definitely need to treat yourself this Valentine's Day![19]

It seems terribly ironic that in an age of instant communication on a variety of fronts (phone calls, emails, texts, chat rooms, dating sites, and Facebook) that people seem lonelier than ever. Somehow the gadgetry that connects us is missing true relationship and leaving our hearts empty.

Paul warned Timothy of the perilous end times with these words, "Men will be lovers of themselves" (2 Timothy 3:2). As we noted in the opening lessons, God made us lovers at heart, but that love is first to be for God above all else, and secondarily for others. The self-absorption of our times shortchanges our love capacitites and is nothing but ruinous to our very humanity.

Processing............➤

1. What cultural signs of self-centeredness are most evident to you?

2. How do you see technology contributing to the breakdown of relationships?

3. What are ways that technology can be used to bless and nurture relationships?

4. What guidelines do you follow to resist allowing technology to intrude on your interactions and conversations with other people?

5. What would be your response if someone were to start a dating site for conservative Anabaptists, and what reasons would you give for your response?

Lesson 9

Sexual Snares

Objectives:
- To realistically assess the increased dangers of sexual sin via digital technology.
- To explore pratical strategies for fostering healthy sexuality and avoiding sexual snares.

○ ○ ○ ○ ○

Trap 6: Sexual Snares

Certainly, one of the most constant and destructive dangers of electronic media is sexual temptation. Pornography is the largest industry on the internet, and pornography sites are the most sophisticated and pernicious in their ability to infiltrate other internet activities. Here are some sobering statistics from a site called webroot.com:

- Every second, 28,258 users are watching pornography on the internet

- Every second, $3,075.64 is being spent on pornography on the internet

- Every second, 372 people are typing the word "adult" into search engines

- 40 million American people regularly visit pornography sites

- 35% of all internet downloads are related to pornography
- 25% of all search engine queries are related to pornography, or about 68 million search queries a day
- One third of pornography viewers are women
- Search engines get 116,000 queries every day related to child pornography
- 34% of internet users have experienced unwanted exposure to pornographic content through ads, pop ups ad, misdirected links, or emails
- 2.5 billion emails sent or received every day contain pornography
- Every 39 minutes, a new pornography video is being created in the United States
- About 200,000 Americans are "pornography addicts"[20]

The same website notes, "The societal costs of pornography are staggering. The financial cost to business productivity in the U.S. alone is estimated at $16.9 billion annually; but the human toll, particularly among our youth and in our families, is far greater." They then quote Dr. Patrick F. Fagan, who declared, "two recent reports... on the pornographic content of phone texting among teenagers, make clear that the digital revolution is being used by younger and younger children to dismantle the barriers that channel sexuality into family life."[21]

In his book *Making Choices: Finding Black and White in a World of Grays*, Peter Kreeft makes the observation that "a remarkable change" has come about in our time. He recognizes that lust has always been a part of human interaction, no matter what the era or culture. "But lust has fundamentally changed its origin in the modern world. Lust used to come from the flesh, from the individual's fallen natural desires; now it comes also from the world, from social conditioning."[22]

We might ask why digital technology is so potent in luring men and women into sin.

A number of factors contribute to the power of digital pornography. First, technology allows people to record and produce high-resolution graphic images and edit them for maximum appeal. What once was unrealistic gratification of self only in the imagination, is now made to appear like real life and readily available. Furthermore, a person is able to pursue this unreal world in privacy and anonymity—no one knows and no one will find out. In this hotbed of unrestrained sin, lust has free rein and there are no apparent consequences.

In reality, of course, this is shot through with delusion. Those who produce pornography care nothing for those who view it, only for the dollars they generate. According to numerous sources, many ladies who perform the sexual acts in pornographic films are doing so against their will as slaves of souteneurs. Furthermore, secret sin will someday come to the light. It is only a matter of time. And eternity. And there are consequences. Minds are being corrupted. Bodies are being wasted. Souls are being destroyed. Manhood and womanhood are being mutilated and ruined.

The society that understands freedom as the right to do as we please and combines that with blatant denial of moral absolutes has a recipe for moral collapse. That is what we are seeing in western society. And the digital world makes it easy for Christians to join this slide to debauchery.

Christians in the Digital Age

Processing `------------->`

1. Consider doing an anonymous survey to determine how prevalent this danger is in your group. Possible questions include:

    a. How frequently do you see titillating articles or advertisements via computer or phone?

    b. What has been the instrument of your greatest temptation? (phone, computer, books, magazines, other)

    c. What digital activity or media creates the most temptation for you? (games, music, news, social sites, digital shopping, pornography sites, texting, other)

    d. Have you ever accessed digital pornography? And if so, was it intentional or unintentional, or both?

    e. Do you struggle with accessing sites that offer sexual stimulation? If yes, approximately how frequently?

    f. Do you use an internet filter? And if so, do you find it helpful?

    g. Are you completely open with someone about your sexual temptations via digital media?

    h. Would you welcome more help in dealing with sexual temptation?

2. What strategies for resisting temptation have you found helpful? What strategies have not been helpful?

3. Make a list of Scriptures that address healthy sexuality.

4. Do you agree or disagree with Kreeft that culturally we face more temptation from the world than in times past?

5. Doing well in a highly sexualized and sexually misguided culture calls for a two-pronged approach. We need to develop strategies to resist temptations to go against the ways of God, and we need to cultivate healthy sexuality. In light of this, consider these questions:

    a. What are the dangers of focusing on only one of these approaches?

    b. What are practical and effective ways to cultivate healthy sexuality?

74

6. Sexual temptations are always based on lies. Identify some of the most commonly believed lies that drive sexual sins.

STRATEGIES TO FLOURISH

← PART IV →

Lesson 10: Assessment

Lesson 11: Connection

Lesson 12: Creativity
            Quiet

Lesson 13: Freedom
            Learning

Introduction to Part IV: Strategies to Flourish

In response to the moral and social upheaval in the 1960s, Francis Schaeffer produced a series of lectures and wrote a book entitled *How Should We Then Live?* The same question confronts us as we think about the challenges and opportunities of the technological age. In light of the values that form the community of faith, given the far-ranging potential of technology, and with an open eye to consider the traps surrounding the use of technology, how indeed should we live?

Our goal should be more than survival. We won't do much for the kingdom of God by focusing all our strategies on defenses, though healthy defense is surely necessary. How can we make spiritual progress in our time? How can we harness the mechanisms of technology to carry forward God's kingdom?

In this section, we will explore six strategies for moving forward in the age of technology.

Lesson 10

Assessment

Objectives:
- To Evaluate Paul's questions for assessing our behaviors as Christians and apply them to our assessment of digital technology.
- To consider practical ways to involve fellow believers in assessing our use of digital technology.

○ ○ ○ ○ ○

Strategy #1: Assessment
Ways to Evaluate the Use of Technology

Whether using familiar technology or facing the new and improved, we need to be continually assessing. Paul's guidance to the Ephesians applies to us in our time: "See then that you walk circumspectly, not as fools but as wise, redeeming the time, because the days are evil" (Ephesians 5:15, 16). The English word "circumspectly" literally means "looking all around" and captures the urgency of the original Greek. We need to be alert in all directions for both dangers and opportunities.

Typically when we use technology, we are assessing, but sometimes only from a particular vantage point. It is easy for a businessman, for example, to assess from the angle of profitability. Is it cost effective? Does it reduce labor and increase productivity? Is it a good investment for the years to come? And so on.

A young person might assess from the standpoint of fascination. Is it exciting? What can it do?

Christians sometimes assess only from the standpoint of morality. Is it right or wrong?

While we certainly should consider such criteria as economic profit, interest level, and moral boundaries, taking one of these criterion by itself is not looking in all directions—is not considering all the issues at stake.

In writing to the Corinthians about food offered to idols, the Apostle Paul said, "All things are lawful for me" (1 Corinthians 10:23). But Paul's assessment didn't stop there. He offered a number of additional ways to think about the issue:

1. Is it helpful (v. 23)?

2. Does it edify (v. 23)?

3. What is the effect on others (vv. 24-30)? Specifically, will it cause anyone to stumble (v. 32)?

4. Does it glorify God (v. 31)?

5. Does it control me (6:12)?[1]

As we apply these questions to the use of technology, a number of guidelines will help us. First, it is necessary to think big-picture, not just about personal or immediate benefit. We must think larger than a specific instrument (such as an electronic tablet) or usage (such as reading). We should consider not only ourselves but also our children, our fellow believers, and unbelievers around us. We should think about the impact of technology on all of life—on our social interactions, on thinking habits, on attention span, on our spiritual vitality, and on the environment.

Second, as we assess, it is good to remind ourselves to think on potentials, not only on dangers. The Christian whose life and resources are committed to God and His purposes will constantly be looking for ways to use technology to help accomplish those purposes. As we noted earlier, this urge to use technology for good has enabled innovative missionaries to put the Bible (and many study resources) on a flash drive or SD card. Bible translators are using computers to speed up the process of Bible translation. The question "Is it helpful?" would certainly help us eliminate time-wasting or mind-numbing uses of technology, but on the positive side, it would urge us to ask, "How can this technology be useful and beneficial in family life, education, business, evangelism, worship, and relationships—considering the whole range of human experience?"

A third guiding principle is to open ourselves to the input of our fellow believers, taking advantage of the wisdom of our local church. Because technology is changing our world so rapidly and has both immediate impact and long-term implications, it is hard to be aware of all the effects. What looks dangerous from one perspective might look helpful and even necessary from another point of view. Fellow members can often see potential for good or ill in our use of technology that we aren't seeing personally. Likely some matters can be left to individual preference (as Paul advised the Romans in the matter of kosher meat or the observance of days), but we do well to bear in mind that we don't always understand ourselves well, or see the full effects of our actions, without the help of our brothers and sisters.

As we assess technology and our uses of technology, we will need to exercise the grace and wisdom of the Holy Spirit. We need to be alert, committed, transparent, and kind to one another. We need to practice open and loving communication. We need to keep our mission in focus. And certainly, we need to live in the profound and heartfelt love of God.

Processing.--------------->

1. Discuss each of the evaluative questions above taken from Paul's words to the Corinthians. Attempt to provide examples of where the use of technology was helpful and unhelpful, edifying and destructive, a means of stumbling and a means of grace, etc.

2. What additional questions (besides the five listed above) might we consider in assessing our use of technology?

3. How might differences in personal experience affect these evaluations? For example, how might one's misuse of a smart phone affect how that person thinks about "Is it helpful?" or "Does it edify?"

4. What are the negative effects of being one-dimensional in assessing the use of technology? For example, considering only profit in business?

5. Can you provide other examples of one-dimensional assessment of technology?

6. Do you agree that the whole brotherhood is necessary for wise assessment? How does a person respond to a brotherhood decision that seems imbalanced or uninformed—for example, a mostly defensive position? What should a businessman do if he is no longer able to compete in the marketplace without the use of technology that his church disallows?

7. What criteria have you personally found helpful for assessing your use of technology?

Lesson 11

Connection

Objectives:
- To explore uses of digital technology that enhance and preserve relationships.
- To explore ways to assess and guard against negative effects of communicating from a distance.

○ ○ ○ ○ ○

Strategy #2: Connection
Using Technology to Enhance Relationships

We are a connected society and relationally impoverished. What a mystifying irony! We can readily access friends and loved ones, businesses and customers, people near and far away, with just a few clicks. We can have hundreds of friends on social media. And we can still be lonely.

A couple of age-old principles of relationships may help us here. Deep relationships take time. Deep relationships require knowing and being known in a wide range of human interaction. And deep relationships call for commitment.

The connectivity offered via electronic communication can be an incredible blessing when our loved ones are at a distance. Missionaries spending time overseas can notify their loves ones of special needs, give regular reports, and stay tuned to home developments. Family members and friends can converse with, encourage, and pray with each other across thousands of miles. Workers can notify supervisors of problems and needs as they develop when the supervisors are not on site. And on the lighter side, married couples can connect through the day, asking questions, updating each other on a change of plans, and sending love notes.

In all of these examples, however, the connections are meaningful because there is much more to the relationship than the transfer of electronic bits. We are more apt to fall into the traps associated with electronic communication if we rely on it too much or allow it to replace other means of communication. How well can we actually learn to know a person through texting and emails? What suffers when we resort to using electronic media to communicate significant messages—messages that stir deep emotive responses?

One study demonstrated that when we are conversing in person, as little as seven percent of our communication may be in our words.[2] The remainder comes through such things as tone of voice, facial expression, and body language. How much we miss when we rest too much of our relationships on words alone!

So we must develop strategies to resist the creeping tendency of electronic communication to replace physical presence where such presence is possible and necessary. Deep relationships are not developed through texting, but where there are deep relationships, electronic communication can be a great asset in both practical and relational ways.

Thinking in the opposite direction now, we must also protect our relationships from interruption. As technology makes it possible to get in touch with "anyone, anytime, anywhere," we grow accustomed to instant access to other people. Instant access can lead to constant interruption. We typically consider it rude to interrupt a conversation. But now we can be in a conversation, in a meeting, in a class, or in a church service and receive a text. What do we do? When two people are in a face-to-face conversation and one stops to check a text, the felt reality is that the third party is more important. We need strategies that protect our real-wolrd time and relationships.

One of the subtle dangers of using technology to communicate from a distance is using that distance to hide. Without physical presence, one can flirt or use foul language or bully others with greater boldness, using the physical distance for a protective cushion. This gives the illusion of being able to get away with more. The unfortunate reality is that it simultaneously increases the distance between one's heart and face—that is, a person who resorts to technological outlets to cross verbal and relational boundaries is cultivating a soul that is more and more disconnected from who he or she is in real life. And this disconnect in the soul puts actual relationships in jeopardy.

A husband who flirts with another man's wife, in other words, is simultaneously being less real with his own wife and undermining his integrity even with the woman he's flirting with. The shocking irony is that it can feel like he's making an intimate connection, when in reality he's destroying his ability to connect in deep and enduring ways. Even as he may be portraying himself (and deceiving himself) as a deep lover, he is actually becoming more and more a betrayer.

If we would maintain relational connection in a world where more and more words are flying long-distance over the networks, we must anchor our relationships in face-to-face communication.

To this point, we've looked mostly at one-on-one connections. Are there ways we can use technology to enhance larger connections?

In an article in Christianity Today, Lauren Hunter urged pastors to consider a number of ways to use technology for the good of the church. Although her audience and her assumptions may not fit conservative Anabaptist congregations exactly, we may be more alike than we first imagine. She believes, for example, that churches should have websites—that people search websites to find a church to visit when they are in a new community. And she suggests that pastors should blog regularly to "elaborate on points mentioned in the sermon, discuss ministry aspirations, hint at goals for the future, and even bring up personal issues to begin a more authentic communication channel within your church."[3]

Hunter's premise is that more and more people are using the internet to get information and to make connections, and that church leaders should take advantage of the opportunities this affords. She notes that software is available to enhance communication on church committees and projects. Also, many younger people are using online banking options instead of carrying cash or checkbooks. Churches, she suggests, should make online giving an easy (and safe) option for members.

As an instructor at Faith Builders Educational Programs, I find myself leading groups of students in service projects. We have found group messaging apps a great way to stay connected, especially when we are split up into small groups. We can alert the group about a task that needs immediate attention, call for additional help, ask for advice about a problem, note schedule changes, and more.

Technology has made such connectivity possible.

Processing ·---------------➤

1. Assess your use of electronic communication:

   a. In my most important relationships, am I relying on electronic communication to replace necessary face-to-face communication? Would those close to me say the same?

   b. Do I text my friends just because I'm restless?

   c. Am I able to be present with my family and friends, or do I allow texts or emails from others to interrupt our conversations?

   d. Have I used electronic communication to express feelings or motives that I would not want to express in person?

   e. Do I project a different persona in my electronic communication than my real self?

2. In what ways do you observe electronic communication shaping other people's relationships positively and negatively?

3. What strategies have you used to protect your relationships and to keep your communication from becoming superficial?

4. How do you see clipped, abbreviated texting language affecting communication and relationships?

5. In what ways have you been grateful for improved connectivity?

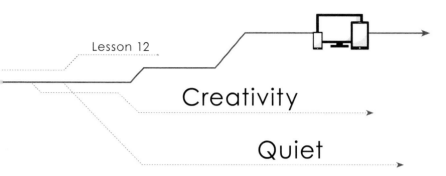

Lesson 12

Creativity

Quiet

Objectives:
- To explore what it means to be creative as God's image-bearers.
- To identify practical ways digital technology can enhance rather than stifle human creativity.
- To explore the value of quietness and consider ways to practice quietness in a noisy age.

○ ○ ○ ○ ○

Strategy #3: Creativity

Pointers for Cultivating Creativity and Ways Technology can Serve

God is endlessly creative. The colors and shapes and sizes and habits and habitats of life forms continue to astound scientists as they explore the world in which we live. The complexity of a simple cell is still beyond us. If we go beyond our world into the universe around us, we are quickly lost, left with our mouths gaping at the enormity "out there." And our God thought it all up and brought it into existence simply by telling it to be. Now He sustains it with His continuing command.

In all this wonder of what is and how it runs, God made humans in His image. And because we are made in His image, we, too, want to make. We can't create in the way He does—we have to start with something—but we are driven to make new things. We invent. We build. We form. We perfect. From handheld tools to the most complex technology, we imagine new things and make them, building on former discoveries and inventions to come up with new ones. In all this imagining and making, we are demonstrating that we are made in the image of the Creator.

Of course, we can use our creativity for good purposes or evil. In the days of Noah, things had gotten so out of hand that God was "grieved" that He even had made humans. "Then the Lord saw that the wickedness of man was great in the earth, and that every intent of the thoughts of his heart was only evil continually. And the Lord was sorry that He had made man on the earth, and He was grieved in His heart" (Genesis 6:5, 6).

So our creativity is a gift both to cultivate and to safeguard.

Television and movies have been part of western culture for more than half a century, making entertainment a central feature of modern life and turning the screen into the most-employed modern babysitter. Educators have attempted to merge entertainment with learning in order to make learning more enjoyable, and digital technology certainly has brought visual and vocal dimensions to documentaries that a textbook cannot. In 1996, the Baby Einstein line of multimedia products hit the market with promises of harnessing the latest technology to maximize the intellectual development of children.

Unfortunately, using the screen to educate seems to result in more steps backward than forward. The overall effects of high doses of digital entertainment have been negative, even when the content is good. In 2011 the American Academy of Pediatrics, after reviewing many studies, issued a policy statement on media for children under two years old, concluding, "there is no such thing as educational TV for this bunch."[4] The report noted, for example, that children develop language skills far better when their parents spend time playing with them than when they watch children's programs on TV.

Other studies have shown that children who spend significant time watching television or movies or playing electronic games have problems with attention span, gain excessive weight, and do not develop analytical thinking. Because the programed media constantly stimulates the emotive centers of the brain, it bypasses the cognitive parts of the brain.[5] Children are attracted to the media, but they aren't exercising their cognitive skills. A sandbox gives them better mental and physical exercise and development than a screen where someone else does amazing things.

Both children and adults have the great opportunity to use and develop their creativity. With new and improved technology continually hitting the market, ought we to be giving more thought to how to harness it for good uses? Is it possible that the same technology the world is using to do its thing could be put to better purposes? And might we sometimes miss those better purposes because we look only at the wrong uses?

Obviously, Christians already have been doing this. But might we do more?

Processing. ------------->

What do you think about the following ideas?

1. Form a technology committee in the local church. Choose to have some younger members who are knowledgeable on the cutting edge of technology and some older members who are anchored in faith and wisdom. Provide them with a scheduled slot in members' meetings to talk about new technology, its uses, abuses, and potential.

2. Make technology reviews a part of mission evaluation. Are there ways technology could be enabling our ministries and mission outreach? Do we equip our missionaries with the same level of technological support we use for our businesses on the home front?

3. Develop a technology statement for use in the local church. What values and guidelines do we follow in our purchase and use of sound equipment, projection equipment, and recording equipment? What protocol do we follow for use of phones and tablets during worship services—devices that have capabilities other than direct church use (Bible programs, pitch instruments, etc.)?

4. Generate and maintain a "creative activities" list for families that posts on the church website. Invite families to offer short reports of creative activities they have done.

5. Similarly, generate a list of documentaries, travel logs, or other media that have educational value for families.

6. Discuss the pros and cons of toys that are designed to simulate people, animals, or machinery (talking, making noises, or performing other functions). How do you think these toys enhance or diminish creativity in children?

Strategy #4: Quiet
Strategies to Preserve Quiet and Rest in a Fast-paced World

Reflecting on silence, Jill Carattini writes,

> Gordon Hempton is of the opinion that you can count on one hand the places in the United States where you can sit for twenty minutes without hearing a generator, a plane, or some other mechanized sound. (His estimation is all the more dreary for Europe.) As an audio ecologist, Hempton has traveled the world for more than twenty-five years searching for silence, measuring the decibels in hundreds of places, and recording the sharp decline of the sounds of nature. "I don't want the absence of sound," he tells one interviewer of his search. "I want the absence of noise." Adding, "Listening is worship."[6]

One of the side effects of modern technology is noise—now referred to as one form of pollution. Responding to this burgeoning problem, nature lovers have created "One Square Inch of Silence" in the remote backwoods of Washington's Olympic National Park. This place is thought to be the quietest spot in the United States, 100% free from noise pollution. The National Park Service regularly monitors the site for sound and takes steps to address any noise intrusion. Visitors have a 3.2 mile walk to the site.

While physical noise is an issue we ought to consider, it reflects the deeper concern of technology's intangible but persistent nudge toward restlessness and motion. This urge is perhaps most discernable in users of smartphones. A study in 2016 found that smartphone users touch their phones more than 2500 times a day, causing Patrick Nelson to conclude:

We're obsessed with our phones.... The heaviest smart-phone users click, tap or swipe on their phone 5,427 times a day, according to researcher Dscout.

That's the top 10 percent of phone users, so one would expect it to be excessive. However, the rest of us still touch the addictive things 2,617 times a day on average.

[Even at night] ...87 percent of 'participants checked their phones, and brought them out of a sleep state, at least once between midnight and 5 a.m.' over the five-day sample period.[7]

How does this incessant urge for stimulation affect our souls? Neil Wiseman suggests that "constant motion" is one of the features of modern life that creates emptiness in the soul. He says that "for spiritually empty people, this ceaseless activity significantly complicates life because it turns each day into a weary rerun of hundreds of unsatisfying yesterdays. Perpetual motion does not drive the dread away or fill the empty soul."[8]

Businesses play music. Waiting rooms feature television sets. And if we prefer our personal options, we can connect earbuds to our phones and choose our own music, get the sound effects of a game we are playing, or tune in to a program without bothering anyone else.

What strategies enable us to use our opportunities well without obliterating all silence and fostering unrest in our souls?

Processing. - - - - - - - - - - - - - - - ➤

What do you think of the following ideas?

1. Use commute time to listen to sermon podcasts or educational talks.

2. Try describing (in a journal) the place of recorded music in your life. How has it shaped you? How has it led your heart? What guides you in your selection of music? Are there ways you would like to change your music habits, and if so, why? How do you intend to do so?

3. Arrange for specific times each day (or longer times once a week) to sit in silence. Perhaps use this time to reflect, meditate, silently worship, or just listen.

4. Limit background music. Choose instead to set aside time to listen to good music where you can focus your attention.

5. Do a log of your electronic activity not related to your work—music, games, texting (by number, not time), social networking sites, news, etc., and then ask evaluative questions:

    a. Do I experience unrest when I don't have anything going?

    b. What activity am I most persistently urged toward when I'm bored?

    c. Am I comfortable with silence and rest?

6. Based on your evaluation, consider a technology fast. Think in terms of further evaluating your ability to enjoy rest and silence.

7. Do an evaluation of your schedule.

    a. Do you regularly feel rushed?

    b. Are you regularly running behind in your schedule?

    c. Do you find electronic devices helping you with scheduling or intruding into your schedule?

    d. How are you most likely to waste time?

    e. If you find yourself continually rushed or behind, how could you plan for margin in your schedule?

    f. If you find yourself continually rushed or behind, what are doing that is not part of your calling? Are there things to which you wish you could say no?

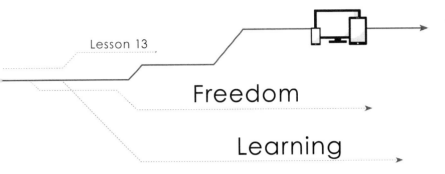

Lesson 13

Freedom

Learning

Objectives:
- To consider biblical principles for breaking sinful addictions and living in freedom.
- To differentiate between information and wisdom.
- To explore guidleines for accessing and utilizing information.

○ ○ ○ ○ ○

Strategy #5: Freedom
Breaking Bondage when Technology has Taken Control

Earlier we recognized that one of the traps of technology is addiction. What strategies can we employ to avoid letting technology control us and to enable us to harness it in good ways? We need to think in two directions here. First, what strategies are helpful for breaking free? And second, what strategies are helpful for getting the most out of our use of technology without becoming controlled by it?

Breaking free of addictive behaviors calls first for accurate appraisal. We can't work with problems we can't look at. And we frustrate our ability to work with a problem to the extent that we frame it inaccurately. To illustrate, a person addicted to alcohol won't take steps to deal with the problem until he can say it is a problem, and if he acknowledges it as a problem but insists it's not a big problem or rationalizes the outcomes, he still frustrates whatever help is available.

So the first step in breaking free of addictive uses of technology is to describe the problem accurately. Typically with addictive behaviors, this takes more than our own assessment. If a young husband spends hours each evening gaming, his wife is likely unhappy. If he says, "I know I probably spend too much time on the computer, but I don't know why she makes it such a big deal; she spends hours having fun with her friends," he may think he has acknowledged his problem. In reality, he has justified it. And in the deceptive light of his justification, he won't feel compelled to make any significant changes.

Consider again the "Ten Signs of Addiction" in Part III and the questions that explore tech addiction. If you want to evaluate your own practices, it can be helpful to write out your responses. Then sit down with someone who knows you well and ask the person to read what you wrote and offer feedback. If you are afraid or unwilling to do that with those who live with you or know you very well, it may mean you are having trouble facing your usage accurately. If discussing your usage (or your responses to the questions) results in significant disagreement and tension, consider inviting a third party to help you assess the issue.

The second thing to address in addictive behaviors is the problem of emptiness. Addictions develop around empty areas in our hearts and then increase the emptiness. We seldom deal adequately with addictions if we simply try to address the unwanted behavior. Almost always, something is driving that behavior. It may be unmet desires. It may be unresolved tension, sorrow, or pain. It may be shallow connection with God. Any of these conditions can make us susceptible to addiction.

Behaviors that stimulate the senses can dull inner tension or pain or can make us oblivious to the deeper hungers of the soul. Especially when sensory satisfaction is quick and easy, it can replace the harder, more disciplined habits that are necessary for spiritual and emotional vitality.

When we learn to use technology to make us feel good, to pacify the deep longing of the human heart, or to dull pain and disappointment, we are using temporary fixes that only increase the problem in the long run. And again, we won't solve the addiction simply by using behaviorist approaches. That is, addictive behaviors don't respond well if our approach is mostly focused on such things as self-discipline, consequences, and accountability meetings.

Assessing our emptiness or our bad habits of filling our emptiness can take time. Most teenagers, for example, don't draw straight lines from their addictions back to their anger at their too-busy parents. Their conscious assumption is typically that they like what they do. Nor does a middle-aged pornography addict easily see how the shame and disillusionment of a shallow and hypocritical lifestyle drives his behaviors.

This brings us to a third consideration in dealing with addictions and the most important one. Breaking free from addiction of any sort calls for the power and presence of Jesus. He is the bondage breaker. He is the life-giver. He is Lord and God. When the Apostle Paul addressed the problem of fleshly lusts and behaviors, he urged, "[P]ut on the Lord Jesus Christ, and make no provision for the flesh, to fulfill its lusts" (Romans 13:14). The Greek wording means, "Clothe yourself with Jesus."

When we attempt to sate the hungers of the heart or salve the wounds of the soul with quick and easy gratification of the flesh, we set ourselves up for addiction. Whatever it is that we turn to soon takes first place in our hearts. We think about it. We pursue it. We love it. We set our goals by it. We sacrifice for it. That "thing" in biblical language is an idol. It does not have life. It cannot give life. It cannot forgive us or redeem us. It cannot make us whole. Nonetheless, it has taken the place that Jesus alone can fill.

But it tastes good and feels good, so it subtly fakes its way into our life rhythms until we are wasted and too weak to quit.

Jesus is our hope.

The best strategy for breaking the bondage of any addiction is a steady and relentless pursuit of Jesus. Without being unpacked, that strategy to many people sounds theoretical and even simplistic. The truth is, however, that Jesus is wisdom. He is strength. He is victory. He is life. And all our strategizing that misses Jesus will miss the way or fall short.

We must learn to know Jesus—clothe ourselves with Him every day—if we wish to break from fulfilling the desires of the flesh and of the mind that are not from God. Those who passionately pursue Jesus see things they never saw before and see old things in new light. Jesus can show us the place technology has in our hearts.

When Jesus has His rightful place in our hearts—when our hearts are aflame with love for Him and we seek His righteousness and His kingdom above all else—we then have a healthy framework for using technology. Even as we jealously guard against dangers surrounding wrong uses, we will explore strategies for using technology for good purposes.

Processing.--------------→

1. What examples can you give of people with addictive behaviors having difficulty naming their problems accurately?

2. Make a list of the benefits of being able to name our sins and problems accurately (see 1 John 1, for starter ideas).

3. Discuss the tie between emptiness and addiction. Does the connection ring true with you? How have you observed this in your own life? How have you seen it in the lives of others?

4. Addictions create weak areas in our lives. How should a believer think of those weak areas after experiencing freedom in Jesus? Consider this scenario: A man has begun texting a lady who is not his wife, leading to an illicit relationship. When this is discovered, he confesses his sin, breaks off the relationship, and sincerely wants to be restored to his wife and family and the church. What guidance does he need in order to move forward? What kind of restrictions will be helpful? What kind of support does he need? What support is necessary for his wife and children?

5. What are appropriate ways to relate to a teenager whose computer or cell phone use seems to be spiritually and relationally damaging?

6. How should church leaders relate to parents of the above teenager who seem confused or passive in relating to their teen?

7. Many Christians push back against the controlling elements of technology by doing "fasts" or setting specific limits on their uses. Here are some examples:

    a. No emailing until after morning devotions

    b. No internet after 10:00 p.m.

    c. One month of no recorded music (or Facebook posts, or YouTube, or ___)

    d. Only x-amount of time on social media

    What restrictions have you found helpful? Or what guidelines have you experienced as liberating?

8. What are some things that restrictions can do/ cannot do in controlling the use of technology?

Strategy #6: Learning
Utilizing Technology to Grow the Heart and Mind

Technology has spawned the information age. This doesn't automatically translate into people being smarter, though it can give us that illusion. What it does mean is that we have unprecedented access to information. With a few taps or clicks we can check the pronunciation and meaning of a new word, check the weather forecast locally or in a distant city, do a conversion from centimeters to inches, find the best treatment for a medical problem, get commentary on a passage of Scripture, or access the best price on anything from a pack of paper towels to an airline ticket. Information is literally always at our fingertips.

We should consider a number of things in order to make the most of all this available information.

First, we need to see the place of information in the larger picture of life. Information can be amazingly useful. It can protect us from scams and costly mistakes. It can help us to make and market and manage with efficiency. It can help us understand and work our way through problems. But information, like money or position or name recognition, can easily be used for wrong reasons. If we are selfish and egotistical, if we are stubborn and argumentative, if we are greedy and calculating, information can be a tool to get our way, put down others, or drive hard bargains. Information is a tool. It does not define who we are; rather, who we are will define how we use information and to what ends. We must not make it out to be more than it really is.

Second, information can be dangerous in excess. It can be dangerous in a number of ways. The Apostle Paul warned, "Knowledge puffs up, but love edifies" (1 Corinthians 8:1). If we become information junkies, we can have answers for all sorts of things, but information doesn't automatically translate into character. In the culture of technology, people easily stand in awe of the person who can do amazing things with technology. And almost across the board, younger ones can wow their elders with their electronic capabilities. This quickly translates into older people feeling stupid and their grandchildren looking smart. But information is not wisdom, and it is a great loss when grandchildren become unhinged from the wisdom of their grandparents.

Another danger associated with the excess of information is simply information overload. Bertram

Gross, a political science professor at Hunter College, coined the term "information overload" in his book *The Managing of Organizations*. He concluded that too much information can actually decrease our ability to make good decisions.[9]

This brings us naturally to another consideration. Not all information is equal. What we read might be true, but it might just as well be biased, misleading, or downright false. The internet has made it possible for anyone to post information, so we can access almost any position on any subject and get support for what we want to believe. With the immensity of information available and the wide range of opinions and ideas, people who go to the internet for information need to learn careful discernment.

Furthermore, some information is not helpful, even when it is true. News sites that focus on the strange or the titillating can bombard readers with information that is simply not necessary to know or worse, that is detrimental to know.

If we are going to do well in an era where we can access anything under the sun, we had better develop strategies of discernment.

Perhaps one of the subtler dangers of information access is in the area of critique. The internet has provided an avenue for virtually anyone to investigate and criticize anyone and to do so with a layer of protection—no physical, face-to-face presence. Certainly, there is a place for critique, for rebuttal, and for rebuke. In the early church, we find Paul confronting Peter (Galatians 2), and Peter rebuking Simon (Acts 8).

But the scriptural approach to confronting a brother has clear and careful boundaries. We are to go to one another in private (Matthew 18:15). We are to be "spiritual" (Galatians 6:1), which means we are to demonstrate mature graces and methods guided by the Holy Spirit, shown especially in humility, approachableness, gentleness, and goodwill. We are to be motivated by love—having a deep and heartfelt commitment to the good of others—and specifically, we are to avoid quarrelsome approaches and tactics (2 Timothy 2:24).

When Christians use public sites to find and spread negative information about fellow believers or to post arguments and criticisms about each other, they easily ignore these directives. The comment sections of many blog posts too quickly descend into harsh judgments and endless arguments.

As noted earlier, there certainly is a place for rebuke, for correction, and for pointing out error. There is also profit in discussion and debate among believers. But Christians should exercise kindness, consideration, and fairness in all situations. This calls us away from evil speaking. It calls us to carefully evaluate information posted about other believers—not only if it is true, but if it ought even to be posted publicly. It calls us to treat others as we would wish to be treated ourselves. Defenders of the faith must not resort to using tactics of the flesh to win arguments and stand for truth.

Information, then, does not automatically translate into wisdom. Knowing more than others does not mean we are superior. With poor character, in fact, information only gives us more potential to do harm rather than good. As James says:

"If you have bitter envy and self-seeking in your hearts, do not boast and lie against the truth. This wisdom does not descend from above, but is earthly, sensual, demonic. For where envy and self-seeking exist, confusion and every evil thing are there. But the wisdom that is from above is first pure, then peaceable, gentle, willing to yield, full of mercy and good fruits, without partiality and without hypocrisy. Now the fruit of righteousness is sown in peace by those who make peace" (James 3:14-18).

The information age has much potential for good among those who love God with all their hearts and their neighbors as themselves. It likewise has much potential for harm for those who value their own opinions above all others, who are self-conceited, who are self-righteous, and who simply want to win arguments.

Processing ╌╌╌╌╌╌╌╌╌╌╌➤

1. List and discuss ways you have benefited from readily available information.

2. How have you observed people using available information wrongly or unwisely?

3. In what ways have you found information on the internet to be misleading?

4. What criteria have you found helpful for evaluating information (not only on the internet but through advertising or promotional literature)?

5. What are the clues that a site is misleading or biased?

6. What are wise responses to people who seem to be relying on misleading or biased information?

7. How should we think about and to what extent should we utilize "watchdog" sites?

⋯⋯ Appendix ⋯⋯➤
Digital Technology Evaluation Tool

So where do we go from here?

As Christians, we are commanded to love God with all our hearts, to love others as ourselves, and to help people from all nations follow Jesus faithfully.

Our discussions about digital technology must navigate within that framework. Whatever practices we incorporate into our daily and weekly rhythms are to lead us toward deeper love for God and others, and are to enable us to make better disciples of Jesus, whether those are our children or our neighbors.

Furthermore, our discussions about digital technology are not ends in themselves. They must lead us toward wise decisions. While it is not the goal here to dictate those decisions, what follows will hopefully highlight the kinds of issues we need to face and the kinds of decisions we need to make.

I. Who decides?

Generally, the church has recognized a balance between corporate decisions and individual decisions. Some issues are basic enough to Christianity or significant enough in a particular setting that they call for a corporate position. The early church decided, for example, that all members were to abstain from foods offered to idols, even though mature believers knew that "meat is meat," no matter what idol it had been dedicated to or who sold it. But some issues are best left for the individual conscience, particularly where different persons, different backgrounds, or different situations could allow for different practices. On matters like Jewish food regulations and observance of holy days, then, the early church encouraged believers to respect differing faith practices. With digital technology, not every individual or group will agree exactly where the balance lies, but certainly, we need to consider the need for balance. And if we are going to be part of a community-minded congregation, we will at times need to yield individual liberties and preferences to group decisions.

II. What criteria do we use?

While the list that follows is not comprehensive, it provides a number of criteria to consider in evaluating digital technology:

1. Morality:
   Does this gadget, program, or activity in any way violate the righteousness of God? Is there any contradiction between this and the character

of Jesus? (Note here that morality is not the only criterion to think about—to ask only if something is right or wrong misses bigger issues like our love for God and our mission in the world.)

2. Effect:

How does this gadget, program, or activity shape its users? How does it affect relationships? Will it have positive or negative effects on my character? Does it refresh me, equip me, engage me, or enable me? On the other hand, does it exhaust me or leave me feeling empty, wasted, or ashamed? Does this contribute to or take away from wholesome interaction with peers, across generations, or between genders? Is it intellectually stimulating or dulling?

3. Influence:

If I use this, how will it affect those around me? Will I be ashamed for others to know that I engage in this? Will it enable or detract from my/our witness to the world and our testimony of Jesus?

4. Benefit/Usefulness:

How does this save/waste time and resources? Is it spiritually, relationally, vocationally, intellectually, and economically worth using? How does it enable or how does it detract from the kingdom of God? What good do we get or could we get from using this device or engaging in this activity?

5. Resources:

How much time, money, and effort does it require

from me, not only immediately but also in an ongoing way? Does my involvement assume on the resources of others? How does this device align with the good use of the earth's resources? Will it soon become a throwaway?

6. Dangers:

   What are the potential misuses? Who is most vulnerable to those misuses? Are there ways to safeguard against the dangers? Are the risks worth the benefit?

7. Sacrifices:

   What will I give up in order to have this gadget or engage in this activity? (Change always involves some loss; when we start doing something differently, we quit doing something else.) Does the gain outweigh the loss?

III. What gadgets, programs, apps, or activities should we talk about?

As new technologies develop, these no doubt will change. Here are a number of current ones:

1. Smart Phones and Tablets

   a. What level of accessibility should we have to the internet?

   b. At what age should children/teens own or have access to cell phones?

   c. What rules of etiquette should guide cell phone use (mealtime, visiting with friends, in public meetings, at church, at social gatherings, at the workplace)?

    d. Will this device open me or my children to relationships with strangers? And if so, is that intentional or inadvertent?

    e. What apps are helpful and consistent with Christian values? What apps pose significant risks?

    f. What texting policies ought Christians to observe? What texting policies are necessary for children/adolescents to observe?

    g. How do parents teach healthy practices to their children?

    h. Are there activities, places, or blocks of time when it would be good to observe a "no cell phone" policy?

2. Internet

    a. What are good blocking and accountability programs?

    b. What are the characteristics of effective accountability groups? What makes them ineffective?

    c. When should parents begin talking to their children about the uses and abuses of the internet?

    d. What are practical tips to avoid letting curiosity take us to bad places?

    e. What are appropriate ground rules for using the internet in the home? In a Christian business? In education?

3. Games

   a. What principles guide/limit the use of electronic games?

   b. What limitations should be placed on children for playing electronic games? And why?

   c. How can adults pattern wise use of electronic entertainment? Should there be different guidelines for adults and children playing electronic games?

4. Videos

   a. What standards guide Christians in watching video clips (for example, on YouTube)? What are distinctions between informational and entertaining videos? And should we make those distinctions?

   b. Is there value in Christians watching what is entertaining?

   c. How can parents monitor what their children watch?

   d. What are the implications of a shift from a verbal orientation in story telling to a visual orientation?

5. Music

   a. How should we think about the use of earpieces when in the presence of others?

   b. What are appropriate ways for parents to stay informed of the listening practices of their children/adolescents?

   c. Should we plan for times of silence? And if so, what is the rationale?

Notes

Part I

1. James K. A. Smith, *Desiring the Kingdom* (Grand Rapids, MI: Baker Academic, 2009), 26, 32, 33.

2. Henry Drummond, *The Greatest Thing in the World*, (Springdale, PA: Whitaker House, 1981), 50.

3. The three points that follow are taken from Chapter 7 of Joseph Hellerman's book, *When the Church Was a Family*, (Nashville, TN: B&H Academic, 2009).

4. On this Hellerman writes, "The thought that one could somehow acquire a 'personal relationship with God' outside the faith family—and remain an 'unchurched Christian'—was simply inconceivable to those whose lives had been defined from birth by the groups to which they belonged" (p. 124).

Part II

1. Alvin Toffler, *Future Shock*, (New York, NY: Bantam Books Inc., 1970), 26.

2. Fell, Andy, "World's First 1,000-Processor Chip" https://www.sciencedaily.com/releases/2016/06/160617215802.htm

3. Op. cit.

4. Gelernter, David, "How Hard Is Chess?" http://content.time.com/time/magazine/article/0,9171,137690,00.html

5. Op. cit.

6. Gallup, George, quoted in Iowa City Press Citizen, April 15, 1953, p. 3, col. 1.

7. Masnick, Mike, "The Sky Is Rising" https://www.techdirt.com/articles/20120129/17272817580/sky-is-rising-entertainment-industry-is-large-growing-not-shrinking.shtml

Part III

1. Holst, Arne. "Number of Smartphone Users in the U.S. 2010-2023." Statista, August 30, 2019. https://www.statista.com/statistics/201182/forecast-of-smartphone-users-in-the-us/. and "How Much Does a Cell Phone Plan Cost? - CostHelper.com." CostHelper. Accessed January 27, 2020. https://electronics.costhelper.com/cell-phone-plans.html.

2. "Internet Pornography by the Numbers; A." Cybersecurity & Threat Intelligence Services. Accessed January 27, 2020. https://www.webroot.com/us/en/home/resources/tips/digital-family-life/internet-pornography-by-the-numbers.

3. Lamkin, Paul. "Smartwatch Sales To Soar... Apparently." Forbes. Forbes Magazine, February 4, 2016. https://www.forbes.com/sites/paullamkin/2016/02/03/smartwatch-sales-to-soar-apparently.

4. Frank, Allegra. "Take a Look at the Average American Gamer in New Survey Findings." Polygon. Polygon, April 29, 2016. https://www.polygon.com/2016/4/29/11539102/gaming-stats-2016-esa-essential-facts.

5. Uzoma, Kay. "How Much TV Does the Average Child Watch Each Day? ." Hello Motherhood. Leaf Group, December 5, 2018. http://www.livestrong.com/article/222032-how-much-tv-does-the-average-child-watch-each-day.

6. Kulman, Randy. "How Much Time Do Kids Spend With Technology?" LearningWorks for Kids, April 20, 2016. http://learningworksforkids.com/2015/07/how-much-time-do-kids-spend-with-technology.

7. Ducharme, Jamie. "Kids Spending More Time on Mobile Devices Than Ever." Time. Time, October 19, 2017. https://time.com/4989275/young-children-tablets-mobile-devices.

8. Entertainment Software Association, "2019 Essential Facts About the Computer and Video Game Industry," https://www.theesa.com/wp-content/uploads/2019/05/ESA_Essential_facts_2019_final.pdf.

9. Eyal, Nir. "Who's Really Addicting You to Technology?" LinkedIn. NirAndFar.com, February 9, 2016. https://www.linkedin.com/pulse/whos-really-addicting-you-technology-nir-eyal?trk=eml-b2_content_ecosystem_digest-hero-14-null&midToken=AQGDgytuBKVDsw&fromEmail=fromEmail&ut=2A2RGux5bG3D81.

10. Rubin, Bonnie Miller. "Young People Spend 7 Hours, 38 Minutes a Day on TV, Video Games, Computer." Los Angeles Times, January 20, 2010. http://articles.latimes.com/2010/jan/20/business/la-fi-youth-media21-2010jan21.

11. Hun, Katie, and Naomi Ng. "Man Dies after 3-Day Internet Gaming Binge." CNN. Cable News Network, January 19, 2015. http://www.cnn.com/2015/01/19/world/taiwan-gamer-death/index.html.

12. Kim Hartman, "Wife sells video game-obsessed husband on Craigslist," Nov 21, 2011, http://www.digitaljournal.com/article/314804#ixzz1eX2vbPNg

13. Note this example: "Pornography triggers brain activity in people with compulsive sexual behaviour – known commonly as sex addiction – similar to that triggered by drugs in the brains of drug addicts, according to a University of Cambridge study published in the journal PLOS ONE." The article goes on to say, however, that "the researchers caution that this does not necessarily mean that pornography itself is addictive." So while there is correlation between these behaviors and brain activity, this doesn't automatically mean causation. http://www.cam.ac.uk/research/news/brain-activity-in-sex-addiction-mirrors-that-of-drug-addiction

14. This list is adapted from the following site: https://betteraddictioncare.com/2017/02/10-signs-and-symptoms-of-addiction/?gclid=CKCvpOvr79QCFRdXDQodT9oPkQ

15. Lowry, Rachel. "Meet the Lonely Japanese Men in Love With Virtual Girlfriends." Time, September 15, 2015. http://time.com/3998563/virtual-love-japan.

16. Lylah M. Alphonse, Boston Globe, Posted March 8, 2010.

17. Mildred Tengbom, *Grief for a Season*, (Minneapolis, MN: Bethany House Publishers, 1989), 34.

18. DAILYMAIL.COM REPORTER, January 27, 2015.

19. "Why You Should Be Your Own Valentine," Katie Medlock, Feb 1, 2016.

20. "Internet Pornography by the Numbers; A." Cybersecurity & Threat Intelligence Services. Accessed January 27, 2020. https://www.webroot.com/us/en/home/resources/tips/digital-family-life/internet-pornography-by-the-numbers.

21. Op. cit.

22. Peter Kreeft, *Making Choices*, (Cincinnati, OH: Servant Books, 1990), 96.

Part IV

1. In 1 Corinthians 6:12, Paul uses identical wording ("All things are lawful for me"), suggesting that this was perhaps a saying being used among the Corinthian Christians. As with many proverbs, it is a truth that, when taken out of context, can lead down wrong paths. Whether we are assessing food or technology, determining that something is "lawful" is inadequate in itself to guide us in its use.

2. This comes from research done by Albert Mehrabian in 1971. His ratio for conveying meaning was 7% words, 38% tone of voice, and 55% body language. These percentages have been quoted extensively through the intervening years. Taken wrongly, of course, they can discount the power of words. Mehrabian noted that the meaning of words decreases according to the disconnect between words and body language.

3. Lauren Hunter, "The Technophobe's Ministry Survival Guide," http://www.christianitytoday.com/pastors/2016/december-web-exclusives/technophobes-ministry-survival-guide.html

4. Rochman, Bonnie. "'Educational TV' for Babies? It Doesn't Exist." Time, October 18, 2011. http://healthland.time.com/2011/10/18/why-educational-tv-for-babies-doesnt-exist/.

5. Contributing to the high stimulation, the average shot length (ASL) in movies is commonly 5-8 seconds. Action movies intentionally decrease shot length to enhance the feel of a fast-paced story, with some movies having an ASL of less than two seconds.

6. Jill Carattini, "Sitting With Silence," *A Slice of Infinity*, April 10, 2017, RZIM

7. Patrick Nelson, NETWORKWORLD, July 7, 2016, http://www.networkworld.com/article/3092446/smartphones/we-touch-our-phones-2617-times-a-day-says-study.html

8. Neil Wiseman, *Growing Your Soul*, (Grand Rapids, MI: Fleming H. Revell, 1996), 22.

9. Bertram Gross, *The Managing of Organizations*, (New York, NY: The Free Press, a division of Simon & Schuster, 1964).